PRE TEST®

High-Yield
Basic Science

Notice

Medicine is an ever-changing science. As new research and clinical experience broaden our knowledge, changes in treatment and drug therapy are required. The authors and the publisher of this work have checked with sources believed to be reliable in their efforts to provide information that is complete and generally in accord with the standards accepted at the time of publication. However, in view of the possibility of human error or changes in medical sciences, neither the authors nor the publisher nor any other party who has been involved in the preparation or publication of this work warrants that the information contained herein is in every respect accurate or complete, and they disclaim all responsibility for any errors or omissions or for the results obtained from use of the information contained in this work. Readers are encouraged to confirm the information contained herein with other sources. For example and in particular, readers are advised to check the product information sheet included in the package of each drug they plan to administer to be certain that the information contained in this work is accurate and that changes have not been made in the recommended dose or in the contraindications for administration. This recommendation is of particular importance in connection with new or infrequently used drugs.

High-Yield
Basic Science

McGraw-Hill
Medical Publishing Division

New York   Chicago   San Francisco   Lisbon   London   Madrid   Mexico City
Milan   New Delhi   San Juan   Seoul   Singapore   Sydney   Toronto

McGraw-Hill

A Division of The McGraw·Hill Companies

PreTest® High-Yield Basic Science

Copyright © 2002 by **The McGraw-Hill Companies,** Inc. All rights reserved. Printed in the United States of America. Except as permitted under the United States Copyright Act of 1976, no part of this publication may be reproduced or distributed in any form or by any means, or stored in a data base or retrieval system, without the prior written permission of the publisher.

1 2 3 4 5 6 7 8 9 0   DOC/DOC   0 9 8 7 6 5 4 3 2 1

ISBN 0-07-138630-0

This book was set in Berkeley by North Market Street Graphics.
The editor was Catherine A. Johnson.
The production supervisor was Phil Galea.
Project management was provided by North Market Street Graphics.
The cover designer was Li Chen Chang / Pinpoint.

This book is printed on acid-free paper.

Library of Congress Cataloging-in-Publication Data

Pretest high-yield basic science.
    p.; cm.
  ISBN 0-07-138630-0 (alk. paper)
  1. Medical sciences—Examinations, questions, etc. 2. Physicians—Licenses—United States—Study-guides.
  [DNLM: 1. Biological Sciences—Examination Questions. 2. Medicine—Examination Questions. WB 18.2 P942 2001]
R834.5 .P733 2001
610'.76—dc21

                               2001044253

Contents

Preface

The current format of the United States Medical Licensing Examination (USMLE) Step 1 emphasizes clinical scenarios as the primary test questions. In order to answer these questions, students must know those key concepts most likely to be tested. *PreTest® High-Yield Basic Science* presents this core information, based on the published content outline for the USMLE Step 1, in a portable format to enable students to prepare for their board examinations in spare moments—on the bus or train, between classes, and anywhere a book can be carried.

The high-yield facts in this book were culled from the nine PreTest® Basic Science Series books. The publisher acknowledges and thanks the following authors for their contributions to this book:

Anatomy, Histology, & Cell Biology: Robert Klein, Ph.D. and
    James C. McKenzie, Ph.D.
Behavioral Sciences: Michael H. Ebert, M.D.
Biochemistry & Genetics: Golder Wilson, M.D.
Microbiology: Richard C. Tilton, Ph.D.
Neuroscience: Alan Siegel, Ph.D. and Heidi Siegel, M.D.
Pathology: Earl Brown, M.D.
Pathophysiology: Maurice Mufson, M.D.
Pharmacology: Arnold Stern, M.D., Ph.D.
Physiology: James C. Ryan, Ph.D. and Michael Wang, Ph.D.

McGraw-Hill
August 2001

High-Yield Facts in Anatomy, Histology, & Cell Biology

Embryology

Embryological development is divided into three periods:

The **Prenatal Period** consists of **gamete formation** and maturation, ending in fertilization.

The **Embryonic Period** begins with fertilization and extends through the **first eight weeks** of development. It includes implantation, germ layer formation, and organogenesis. This is the critical period for susceptibility to **teratogens**.

The **Fetal Period** extends from the **third month** through birth.

THE PRENATAL PERIOD

The **development of gametes** begins with the duplication of chromosomal DNA followed by two cycles of nuclear and cell division (**meiosis**).

Genetic variability is assured by **crossing over** of DNA and by **random assortment** of chromosomes during the first meiotic division. Errors can result in duplication or deletion of all or part of a specific chromosome.

Spermatogenesis

During puberty, primordial stem cells in the walls of the seminiferous tubules of the testes undergo mitotic divisions to replenish their population and form a group of **spermatogonia** that will undergo meiosis.

Primary spermatocytes are spermatogonia that have duplicated their DNA (4N).

Secondary spermatocytes result from the first meiotic division (2N).

Spermatids are formed by the second meiotic division (1N).

Spermiogenesis

During this phase, spermatids mature into sperm by losing extraneous cytoplasm and developing a head region consisting of an **acrosome** (giant lysosome) surrounding the nuclear material.

The processes of spermatogenesis and spermiogenesis are **continuous** and last about two months.

Oogenesis

Oogenesis begins in the fetal period in females and is a **discontinuous** process involving both meiosis and maturation.

Oogonia form **primary oocytes** but stop in the metaphase of the first meiotic division until puberty. The second meiotic division is not concluded until fertilization occurs. Maturational events include retention of protein synthetic machinery in the surviving oocyte, formation of **cortical granules** that participate in events at fertilization, and development of a protective glycoprotein coat, the **zona pellucida.**

Fertilization

Fertilization occurs when sperm and oocyte cell membranes fuse and the male pronucleus is injected into the oocyte. Following coitus, exposure of sperm to the environment of the female reproductive tract causes **capacitation**, removal of surface glycoproteins from the sperm membrane enabling fertilization to occur.

Binding of the first sperm initiates the **zona reaction**. Release of **cortical granules** causes biochemical changes in the zona pellucida and oocyte membrane that prevent **polyspermy.**

EMBRYONIC DEVELOPMENT

The embryo forms one **germ layer** during each of the first three weeks.

During the second week, the **embryoblast** differentiates into two germ layers, the **epiblast** and the **hypoblast**. This establishes the dorsal (epiblast)–ventral (hypoblast) body axis.

During the third week, the process of **gastrulation** occurs by which epiblast cells migrate toward the **primitive streak** and ingress to form the **endoderm** and **mesoderm** germ layers below the remaining epiblast cells (**ectoderm**).

Lateral body folding at the end of the third week causes the germ layers to form three concentric tubes with the innermost layer being the endoderm, the mesoderm in the middle, and the ectoderm on the surface.

GERM LAYER DERIVATIVES

Mesoderm Derivatives

The mesoderm is divided into four regions (from medial to lateral): axial, paraxial, intermediate, and lateral plate.

Axial mesoderm is midline and forms the notochord.

Paraxial mesoderm forms somites. Somites are divided into **sclerotomes** (bone formation), **myotomes** (muscle precursors), and **dermatomes** (dermis).

Intermediate mesoderm gives rise to components of the genitourinary system.

Germ Layer Derivatives

Ectoderm Derivatives	Epithelium of skin (superficial epidermis layer)	

	All nervous tissue: formed by neuroectoderm:
	  Brain and spinal cord (neural tube)
	  Peripheral nerves and other neural crest derivatives	

Endoderm Derivatives	Epithelial linings of:	The gastrointestinal tract

		Organs that form as buds from
		  the endodermal tube:
		  Pharyngeal gland derivatives*
		  Respiratory system
		  Digestive organs (liver, pancreas)
		  Terminal part of urogenital systems

	Hypoblast Endoderm: Gametes migrate to gonads	
Mesoderm Derivatives	All connective tissues**	General connective tissues
		Cartilage and bone
		Blood cells (red and white)
	All muscle types:	Cardiac, skeletal, smooth
	Epithelial linings of:	Body cavities

		Some organs:
		  Cardiovascular system
		  Reproductive and urinary
		  systems (most parts)

*Pharyngeal derivatives: palatine tonsils, thymus, thyroid, parathyroids
**Some connective tissues in the head are derived from neural crest

Lateral plate mesoderm forms bones and connective tissues of the limbs and limb girdles (**somatic layer**) and the smooth muscle lining viscera and the serosae of body cavities (**splanchnic layer**).

Intermediate mesoderm is *not found* in the head region, and the lateral plate mesoderm is *not divided* into layers there.

Ectoderm Derivatives

Formation of the primitive central nervous system is induced in the ectoderm layer by cells forming the **notochord** in the underlying mesoderm.

The neural plate ectoderm (**neuroectoderm**) forms two lateral folds that meet and fuse in the midline to form the neural tube (**neurulation**).

Cells from the tips of the folds (**neural crest**) migrate throughout the body to form many derivatives including the peripheral nervous system.

FORMATION OF THE HEAD REGION

Neural crest contributes significantly to formation of connective tissue elements in the head.

The bony skeleton of the head is comprised of the **viscerocranium** and the **neurocranium.**

The neurocranium (cranial vault) is composed of a base formed by **endochondral ossification** (chondrocranium) and sides and roof bones formed by **intramembranous ossification.**

The chondrocranium is derived from both **somitic mesoderm** (occipital) and neural crest.

The viscerocranium (face) is derived from the first two **pharyngeal (branchial) arches** (neural crest).

LIMB FORMATION

The limbs form as ventrolateral buds under the mutual induction of ectoderm [**apical ectodermal ridge (AER)**] and underlying mesoderm beginning in the fifth week. *The AER influences proximal-distal development.*

Somatic lateral plate mesoderm forms the bony and connective tissue elements of the limbs and limb girdles while skeletal muscle of the appendages is derived from somites.

Cranio-caudal polarity is determined by specialized mesoderm cells [**zone of polarizing activity (ZPA)**] that release inducing signals such as **retinoic acid.**

Homeobox genes are the targets of induction signals.

Rotation of the limb buds establishes the position of the joints, the location of muscle groups, and the pattern of sensory innervation.

MATURATION OF THE CENTRAL NERVOUS SYSTEM

Both neurons and glia develop from the original neurectoderm forming the neural tube.

Microglia *are the exception:* they develop from the monocyte-macrophage lineage of mesodermal (bone marrow) origin and migrate into the CNS.

Induction of regional differences in the developing CNS is regulated by **retinoic acid** (vitamin A). Overexposure of the cranial region to retinoic acid can result in "caudalization," i.e., development more similar to the spinal cord.

During development, the spinal cord and presumptive brainstem develop three layers: (1) a germinal layer or **ventricular zone,** (2) an **intermediate layer** containing neuroblasts and comprising gray matter, and (3) a **marginal zone** containing myelinated fibers (white matter).

Other layers are added in the cerebrum and cerebellum by cell migration along glial scaffolds.

The notochord induces the establishment of **dorsal-ventral polarity** in the neural tube. Ventral portions of the tube will become the **basal plate** and give rise to motor neurons, whereas the dorsal portions become the **alar plates** and subserve sensory functions.

Meninges are formed by mesoderm surrounding the neural tube with contributions to the arachnoid and pia from neural crest.

Defects in the CNS may result from several causes including high maternal blood glucose levels and vitamin A overexposure and often involve bony defects (e.g., **spina bifida** and **anencephaly**). Defects are most common in the regions of **neuropore** closure.

PERIPHERAL NERVOUS SYSTEM

Sensory neurons of the spinal ganglia, as well as autonomic postganglionic neurons and their supporting cells, are derived from neural crest.

Focal deficiencies in neural crest cell migration may result in lack of innervation to specific organs or parts of organs. In **Hirschsprung disease,** failure of neural crest cells to migrate to a portion of the colon results in a localized deficiency in parasympathetic intramural ganglia that may cause a loss of peristalsis and fatal bowel obstruction.

DEVELOPMENT OF THE HEAD AND NECK

The cartilages, bones, and blood vessels of the face (viscerocranium) develop from the **pharyngeal (branchial) arches.** Each arch receives its blood supply from a specific aortic arch and its innervation from a specific cranial nerve. The **skeletal muscles** of the head and neck primarily arise from the pharyngeal arches and have a unique innervation (special visceral efferent).

The face develops from a midline **frontonasal prominence** and bilateral **maxillary and mandibular prominences.** Failure of the prominences to fuse results in various facial malformations.

Pouch and cleft 1:	Epithelial lining of middle and outer ear canals and tympanic membrane
Pouch 2:	Epithelial lining of palatine tonsils
Pouch and cleft 3:	
Ventral Portion:	Epithelial components of thymus gland
Pouch 3: Dorsal Portion:	Epithelial cells of inferior parathyroid glands
Pouch 4: Ventral Portion:	Epithelial parafollicular cells (incorporate into thyroid gland)
Dorsal Portion:	Epithelial cells of superior parathyroid glands
Clefts 2 and 4:	No derivatives

Teeth originate from both ectodermal (enamel) and neurectodermal (neural crest: dentin, pulp, cementum, and periodontal ligament) derivatives.

DERIVATIVES OF PHARYNGEAL POUCHES AND CLEFTS

The anterior portion of the pituitary is derived from oral ectoderm arising from the roof of the oral cavity (Rathke's pouch) anterior to the buccopharyngeal membrane and migrating through the sphenoid anlagen to unite with a downgrowth (posterior pituitary) from the hypothalamus.

The eye is derived from three different germ layers:

Neurectoderm: Vesicular outgrowths of the forebrain differentiate into **retina and optic nerve.**

Surface ectoderm: Contributes to the **lens, cornea,** and epthelial coverings of the lacrimal glands, eyelids, and **conjunctiva.**

Mesoderm: The **sclera** and **choroid** are derived from lateral plate mesoderm.

The **extraocular muscles** are formed by myotomes of **cranial somitomeres.**

Structures of the **outer and middle ear** are derived from the first and second **pharyngeal arches** and the **first pharyngeal cleft.**

Structures of the **inner ear** are derived from the **ectodermal otic placode.**

Maternal rubella can cause defects in both eye (fourth to sixth weeks of gestation) and ear (seventh to 8th weeks).

FORMATION OF THE CARDIOVASCULAR SYSTEM

All components of the cardiovascular system, including the epithelia, are derived from **splanchnic lateral plate mesoderm.**

The heart tubes forming on either side of the endodermal tube are brought together by **lateral body folding.**

Looping of the heart tube occurs while the tube is being divided into left and right portions by interatrial and interventricular septa.

In the interatrial septum, the **septum primum** and **septum secundum** do not close off the **foramen ovale** until birth.

Failure of the **atrioventricular endocardial cushions** to fuse can result in septal and valve defects.

Neural crest cells contribute to septation of the truncus arteriosus and the formation of the aortic and pulmonary outflows, as well as the aortic arches.

The "**Tetralogy of Fallot**" is the most common defect of the conus arteriosus/truncus arteriosus and involves stenosis of the pulmonary trunk, ventricular septal defect, right ventricular hypertrophy, and overriding aorta.

Vasculature

The endothelial lining of most blood vessels forms by coalescence of **mesodermal cells** and subsequent **vacuolization** to form a lumen. Subsequently, smooth muscle cells and connective tissue elements are supplied by local mesoderm.

The paired doral aortae and the five aortic arches form an early symmetric arterial system. Regression of portions of these vessels later results in the asymmetrical adult arterial system. The **vitelline arteries** connect the yolk sac to the abdominal dorsal aorta. They will form the arteries of the GI tract: **celiac, superior mesenteric, and inferior mesenteric.**

Blood islands formed during etiology of the vitelline arteries are the first sites of **hematopoiesis** and seed other hematopoietic tissues.

The paired **umbilical arteries** develop from the caudal end of the dorsal aorta and invade the mesoderm of the placenta. They carry deoxygenated blood from the fetus to the placenta.

The **caval venous system** is derived mostly from the right anterior and posterior **cardinal veins.**

The **vitelline veins** form the veins of the digestive system, including the **portal vein,** and the terminal part of the inferior vena cava.

No components of the **umbilical veins** remain patent after closure of the ductus venosus.

DEVELOPMENT OF THE HEMATOPOIETIC SYSTEM

Onset of **hematopoiesis** begins with formation of **blood islands** in the wall of the yolk sac (derived from the hypoblast) during week 3.

Pluripotent stem cells from the blood islands seed the other hematopoietic sites. These are, in succession, the **liver** (week 5), **spleen** (week 5), and **bone marrow** (month 6).

All components of hematopoietic organs are derived from **mesoderm** except for the **epithelium of the thymus,** which is derived from endoderm of the **third pharyngeal pouch.**

DEVELOPMENT OF THE DIGESTIVE SYSTEM

The epithelium of the digestive tract and associated organs is formed by the **endoderm tube,** whereas connective tissue and smooth muscle are derived from **splanchnic lateral plate mesoderm.** The mesoderm induces regional specialization in the endoderm.

The midgut endoderm is the last to fold into a tube and remains connected to the yolk sac via the yolk stalk.

Formation of the mesodermal **urorectal septum** divides the cloaca into the **urogenital sinus** and **primitive rectum.**

Cell proliferation results in closure of the endodermal tube lumen during week 6. The lumen is reopened by **recanalization** in week 8.

Failure to recanalize can result in **stenosis,** preventing the passage of amniotic fluid swallowed by the fetus (**polyhydramnios**).

Peristalsis begins in week 10 when cells of neural crest origin invade the muscular layer to form the enteric nervous (autonomic) system. Failure of neural crest cell migration to the distal hindgut results in **aganglionic megacolon (Hirschsprung disease**), which may cause fatal intestinal obstruction.

The adult pattern of GI organ distribution is achieved by **physiologic herniation** and then retraction of the midgut during the second month.

Failure of the midgut loop to return to the abdominal cavity may result in an **omphalocele or umbilical hernia.**

Associated digestive organs (liver, gallbladder, and pancreas) originate as outgrowths of the endodermal tube. Connective tissue components of the liver are derived from both splanchnic and somatic (septum transversum) lateral plate mesoderm. Lateral plate mesoderm also forms the peritoneum and mesenteries of the abdominal cavity.

FORMATION OF THE RESPIRATORY SYSTEM

The first part of the respiratory system is lined by ectoderm derived from the nasal **ectodermal placodes.**

In the fourth week, a **respiratory diverticulum** arises as an outgrowth of the ventral **endodermal tube.**

Endoderm will form the respiratory epithelium, whereas **splanchnic lateral plate mesoderm** will form connective tissue elements including cartilage, smooth muscle, and blood vessels.

Mesoderm directs the branching pattern of the developing airways.

Although most alveoli do not form until after birth, the lungs are capable of sufficient gas exchange after 6.5 months' gestation. **Respiratory distress syndrome** develops in premature births if **surfactant** levels are inadequate.

Abnormal septation of the trachea and esophagus can result in stenosis, atresia, or tracheoesophageal fistulas.

DEVELOPMENT OF THE URINARY SYSTEM

Epithelial structures of the urinary system are derived from two sources: **intermediate mesoderm and urogenital sinus endoderm.**

Three pairs of kidneys develop in cranio-caudal sequence in the urogenital ridge of intermediate mesoderm: **pronephros, mesonephros, and metanephros.**

The caudal end of the mesonephric duct gives rise to the **ureteric bud.** The ureteric bud induces surrounding intermediate mesoderm to form the

metanephric cap, which forms the excretory units of the kidney. The ureteric bud will form the collecting ducts.

The epithelial lining (transitional epithelium) of the **ureters,** as well as their muscular and connective tissue components, are derived from **intermediate mesoderm.** The transitional epithelium of the **bladder** and most of the **urethra** (transitional) is derived from hingut **endoderm of the urogenital sinus.** Connective tissue and muscle are derived from splanchnic lateral plate mesoderm.

DEVELOPMENT OF THE REPRODUCTIVE SYSTEMS

Intermediate mesoderm forms the epithelia, connective tissues, and smooth muscle of the indifferent **sex cords** and their ducts.

The **endoderm of the urogenital sinus** gives rise to the epithelia of distal organs of the reproductive system and the external genitalia. As in the urinary system, connective tissue and smooth muscle of these terminal elements is provided by splanchnic lateral plate mesoderm.

Germ cells migrate from their origins in yolk sac endoderm into the indifferent sex cords of the **urogenital ridge** by week 6. Further differentiation of both the immature sex cords and the germ cells depends on **mutual induction.**

The *Sry* **gene** on the **Y chromosome** directs the differentiation of the medullary sex cords into **testes.** If this gene (or a Y chromosome) is not present, the cortical sex cords will develop as ovaries.

Production of **Müllerian inhibiting substance** by Sertoli cells induces presumptive Leydig cells to produce **testosterone** and other sex hormones that regulate further **male differentiation.**

In the absence of testosterone, developing **follicular cells** of the ovaries direct the **differentiation** of germ cells into **oogonia.**

Two pairs of genital ducts develop in both sexes. **Wolffian (mesonephric)** ducts develop first as part of the urinary system.

Paramesonephric (Müllerian) ducts develop next and are open to the pelvic cavity at their cranial ends. The mesonephric system will persist in the male and the paramesonephric system in the female.

In **males,** the **urogenital sinus endoderm** gives rise to the epithelia of the **urethra** and associated **prostate and bulbourethral glands.**

In the **female,** the endoderm of the **urogenital sinus** is the origin of the epithelium of the **lower vagina,** the upper portion being formed by the **paramesonephric ducts.**

Male differentiation of external genitalia requires testosterone. Female differentiation requires **placental estrogen.**

DEVELOPMENT OF THE PLACENTA
AND FETAL MEMBRANES

The fetal portion of the placenta forms from the **trophoblast.**

Syncytiotrophoblast cells are in direct contact with maternal tissue, whereas the embryo proper is separated from the **cytotrophoblast** by **extraembryonic mesoderm** (together, **the chorion**).

Primary villus: syncytiotrophoblast with a cytotrophoblast core.

Secondary villus: Cytotrophoblast core invaded by extraembryonic mesoderm.

Tertiary villus: Fetal blood vessels invade the mesoderm (week 3).

The presumptive **umbilical blood vessels** form in the wall of the **allantois,** an endodermal outpocket of the urogenital sinus.

The **amnionic membrane** develops from **epiblast** and is continuous with embryonic ectoderm. The lining of the **yolk sac** develops from **hypoblast** and is continuous with embryonic endoderm.

The yolk sac gives rise to the first **blood islands** that will form the **vitelline vessels.**

Passive immunity is transfered to the fetus by diffusion of **immunoglobulin G** from the maternal to the fetal circulation.

Excess amniotic fluid is swallowed by the fetus, absorbed by the fetal GI tract, transferred to the fetal circulation, and finally crosses the placental membranes to the maternal circulation.

Hormones secreted by the placenta include **chorionic gonadotropin,** estrogen, progesterone, and **chorionic somatostatin** (placental lactogen).

Histology and Cell Biology

CELL MEMBRANES

Cell membranes consist of a **lipid bilayer** and associated proteins and carbohydrates. In the bilayer, the hydrophilic portions of the lipids are arranged on the external and cytosolic surfaces, and the **hydrophobic tails** are located in the interior. **Transmembrane proteins** are anchored to the core of the bilayer by their hydrophobic regions and can be removed only by detergents that disrupt the bilayer. **Peripheral membrane proteins** are attached to the surface of the membrane by weak electrostatic forces and are easy to remove by altering the pH or ionic strength of their environment.

CYTOPLASM AND ORGANELLES

Cytoplasm is a dynamic fluid environment bounded by the cell membrane. It contains various membrane-bound organelles, nonmembranous structures (such as lipid droplets, glycogen, and pigment granules), and structural or cytoskeletal proteins in either a soluble or insoluble form. The **endoplasmic reticulum** (ER) is a continuous tubular meshwork that may be either smooth (SER) or rough (RER) where studded with ribosomes. The discoid stacks of the **Golgi apparatus** are involved in packaging and routing proteins for export or delivery to other organelles, including lysosomes and peroxisomes. **Lysosomes** degrade intracellular and imported debris, and **peroxisomes** oxidize a variety of substrates, including alcohol. Only the **nucleus**, which is the repository of genetic information stored in deoxyribonucleic acid (DNA), and the **mitochondria**, which are the storage sites of energy for cellular function in the form of adenosine triphosphate (**ATP**), are enclosed in a double membrane. Also included in the cytoplasm are three proteins that form the **cytoskeletal infrastructure: actin bundles** that determine the shape of the cell; **intermediate filaments** that stabilize the cell membrane and cytoplasmic contents; and **microtubules** (**tubulin**), which use molecular motors to move organelles within the cell.

NUCLEUS

The nucleus consists of a nuclear envelope that is continuous with the ER, chromatin, matrix, and a nucleolus rich in ribosomal ribonucleic acid (rRNA). The **nuclear envelope** contains pores for bidirectional transport and is supported by intermediate filament proteins, the **lamins**. **Chromatin** consists of euchromatin, which is an open form of DNA that is actively transcribed, and heterochromatin that is quiescent. During **cell division,** DNA is accurately replicated and divided equally between two daughter nuclei. Equal distribution of chromosomes is accomplished by the microtubules of the mitotic spindle. The separation of cytoplasm (**cytokinesis**) occurs through the action of an **actin contractile ring.** The cell cycle consists of **interphase (G_1, S, G_2, and M), prophase, prometaphase, metaphase, anaphase,** and **telophase.** The cell cycle is regulated at the G_1/S and G_2/M boundaries by phosphorylation of complexes of a protein kinase [cyclin-dependent kinase (Cdk) protein] and a **cyclin (cytoplasmic oscillator).**

INTRACELLULAR TRAFFICKING

The key event in exocytosis is translocation of newly synthesized protein into the cisternal space of the rough ER (**signal hypothesis**). Proteins and lipids reach the Golgi apparatus by vesicular transport. Using carbohydrate-sorting signals, proteins are sorted from the *trans*-face of the Golgi apparatus to secretory vesicles, the cell membrane, and **lysosomes.** Lysosomal enzymes are sorted by using a **mannose-6-phosphate** signal recognized by a receptor on the lysosomal membrane. Nuclear and mitochondrial-sorting signals (positively charged amino acid sequences) are recognized by those organelles.

Endocytosis involves transport from the cell membrane to lysosomes using endosome intermediates. The process originates with a **clathrin-coated pit** that invaginates to form a **coated vesicle** that fuses with an endosome. This internalization can be **receptor-mediated** (e.g., uptake of cholesterol). Endosomes subsequently fuse with lysosomes. Internalized receptor/ligand complexes may be conserved, degraded, or recycled.

EPITHELIUM

Epithelial cells line the free external and internal surfaces of the body. Epithelia have a paucity of intercellular substance and are interconnected by **junctional complexes.** Components of the junctional complex include the *zonula occludens* (tight junction), which prevents leakage between the adjoining cells and maintains apical/basolateral polarity; *zonula adherens,* which links the actin networks within adjacent cells; and *macula adherens* (desmosome), which links the intermediate filament networks of adjacent cells. Epithelial cells also form a firm attachment to the basal lamina, which they secrete. **Gap junctions** or **nexi** permit passage of small molecules directly between cells. Apical specializations are prominent in epithelia and include microvilli that increase surface area; stereocilia, which are nonmotile modified microvilli; and cilia and flagella, which are motile structures. **Cilia** and **flagella** have the classic "9 + 2" microtubular arrangement emanating from basal bodies.

CONNECTIVE TISSUE

Connective tissue consists of cells and a matrix (fibers and ground substance). The cells include **fibroblasts** (the source of collagen and other fibers), plasma cells (the source of antibodies), **macrophages** (the cells responsible for phagocytosis), **mast cells** (the source of heparin and histamine), and a variety of transient blood cells. **Type I collagen** and elastin make up the predominant fibers found in connective tissue. Ground substance includes proteoglycans and glycoproteins that organize and stabilize the fibrillar network. **Type II collagen** is associated with hyaline cartilage, **type III collagen** forms the collagenous component of reticular connective tissue found in lymphoid organs, and **type IV collagen** forms a sheetlike meshwork of the basal lamina. Other types of collagen exist and include the fibril-associated collagens with interrupted triple helices (**FACIT**). Collagen fibrils are connected to other extracellular matrix molecules by the FACIT collagens.

SPECIALIZED CONNECTIVE TISSUES: BONE AND CARTILAGE

Bone contains three major cell types: **osteoblasts** that secrete type I collagen and noncollagenous proteins; **osteocytes**, which maintain mature bone; and **osteoclasts**, which resorb bone by acidification. Bone deposition is regulated primarily by **parathyroid hormone** (PTH), which is secreted in response to low serum calcium levels. **Calcitonin** opposes the actions of PTH, but plays a lesser role overall. Bone is highly vascular and mineralized with **hydroxyapatite**.

In contrast, the three types of cartilage are avascular and contain chondrocytes that synthesize fibers and ground substance. **Hyaline cartilage** covers articular surfaces and forms the cartilage model in long bone development. **Elastic cartilage** is found in the pinna of the ear and in the larynx, while **fibrocartilage** is an intermediate form found in the intervertebral disc, pubic symphysis, and connecting tendon and bone.

MUSCLE AND CELL MOTILITY

Skeletal and cardiac (striated) muscle contract by sliding **myosin** and **actin** filaments past each other in a process facilitated by ATP. Myosin contains a motor that interacts with the actin filament and allows myosin to ratchet along the actin. The filaments are arranged in a banded pattern in individual **sarcomeres**, which act in series. Specialized invaginations of the plasma membrane (**T tubules**) spread the surface depolarization to the interior of the cell to release calcium from the **sarcoplasmic reticulum**, initiating contraction. **Troponin** and **tropomyosin** are specialized proteins that permit contraction of skeletal and cardiac muscle to be regulated by calcium. Skeletal muscle is a syncytium, while cardiac muscle consists of individual cells connected by intercalated disks. The organization of striated muscle is shown below:

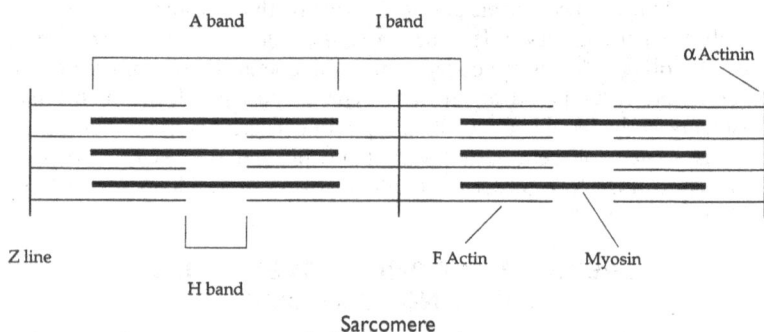

Sarcomere

Smooth muscle contraction closely resembles the cell motility exhibited in other cell types. It also occurs through the action of actin and myosin, which are arranged in a lattice-like pattern.

NERVOUS SYSTEM

Myelin, which insulates neuronal projections and permits rapid (saltatory) conduction, is produced by **oligodendroglia** in the central nervous system (**CNS**) and by **Schwann cells** in the peripheral nervous system (**PNS**). **Microglia** are the macrophages of the brain. **Astrocytes** have a complicated role in physical and metabolic support of neurons. **Neurons** conduct electrochemical impulses and move neurotransmitters to their synaptic termini by **axoplasmic transport**. Transneuronal transmission is accomplished by calcium-regulated release of **synaptic vesicles**. Neurons also synapse with muscle cells. A typical contact between a myelinated neuron and skeletal muscle (**neuromuscular** junction) is shown below.

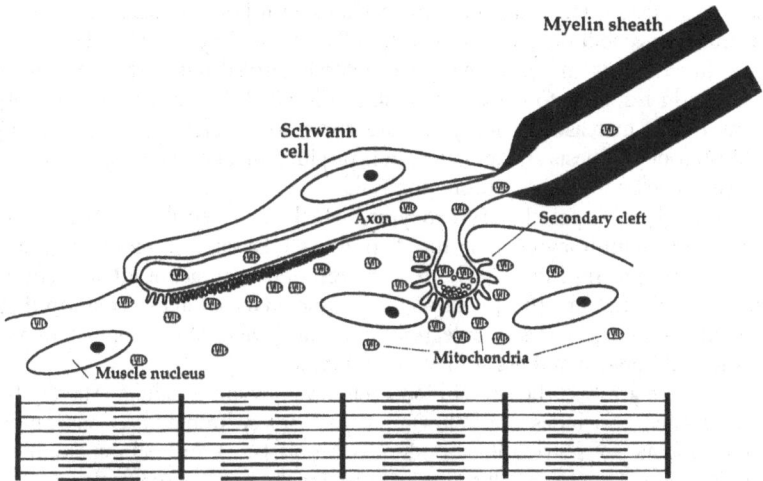

Neuromuscular junction. Axonal terminals (telodendria) rest in shallow depressions (primary clefts) on the surface of the striated muscle fiber. Secondary clefts increase the surface area for interaction with a neurotransmitter (acetylcholine). Muscle cell nuclei and mitochondria are abundant near the junction.

In the **cerebral** and **cerebellar cortices, gray matter** (cell bodies and immediately adjacent processes) is located peripherally and **white matter** centrally; this pattern is reversed in the **spinal cord.** Cerebellar cortex consists of **molecular, Purkinje,** and **granular layers** with extensive arborization of the neuronal processes. The cerebral cortex consists of a homogenous **layer I** with multiple deeper layers of large **pyramidal** and other types of **neurons.** The number of layers varies, depending on the cortical region. Neuronal cell bodies (**perikarya**) are also localized in ganglia in the peripheral nervous system and autonomic nervous system (**ANS**).

CARDIOVASCULAR SYSTEM, BLOOD, AND BONE MARROW

In large arteries close to the heart, the *tunica media* contains high amounts of elastin to buffer the heart's pulsatile output. Smaller muscular arteries distribute blood to organs and capillary beds; their contractions are mediated by both the **sympathetic nervous system** (**SNS**) and by humoral factors. **Endothelial cells** lining the vascular lumen secrete vasoactive substances (e.g., endothelin) and factors important in **blood clotting** (e.g., von Willebrand factor). Smooth muscle cells undergo **hyperplasia** and **hypertrophy** in hypertension. The heart contains specialized cardiomyocytes that function as impulse-generating and conducting cells regulated by the ANS. The heart also functions as an endocrine organ, releasing **atrial natriuretic peptides** (ANPs) in response to increased plasma volume. ANPs reduce plasma volume by (1) increasing urinary sodium and water excretion, (2) inhibiting aldosterone synthesis and angiotensin II production, and (3) inhibiting vasopressin release from the neurohypophysis.

Blood cells include **erythrocytes**, which are specialized for oxygen transport; **lymphocytes** that function in cellular and humoral immune responses; **neutrophils**, which are early responders to acute inflammation; **monocytes** that are the precursors of tissue **macrophages**; **eosinophils**, which respond to parasitic infection; and **basophils**, which contain **histamine** and **heparin** and assist mast cell function.

Bone marrow is the site of blood cell development in adults. The erythrocyte lineage includes the following stages: proerythroblasts → basophilic erythroblasts → polychromatophilic erythroblasts → orthochromatophilic erythrocytes. The white cell series includes myeloblasts → promyelocytes → myelocytes → metamyelocytes → mature granular leukocytes.

LYMPHOID SYSTEM AND CELLULAR IMMUNOLOGY

Functional cells include **B lymphocytes** (humoral immunity), **T lymphocytes** (cellular immunity), **macrophages** (phagocytic and **antigen-presenting cells**), and **mast cells**. Lymphoid organs may be either primary (**bone marrow** and **thymus**) or secondary (**lymph nodes** and dispersed **lymphatic nodules, spleen**, and **tonsils**). The B lymphocytes are educated in the bone marrow and are seeded to specific **B cell regions** of the secondary lymphoid organs, and T lymphocytes are educated in the thymus and seeded to **T cell-dependent regions** of the secondary lymphoid organs. The thymus is recognized by lobulation, separate cortex and medulla in each lobule, the absence of germinal centers, and the presence of Hassall's corpuscles. The lymph nodes, which **filter lymph not blood**, are characterized by a central medulla and a cortex containing primary and secondary follicles. The spleen, which **filters blood**, is characterized by red and white pulp. The **tonsils** are characterized by an **epithelial lining** on one side.

RESPIRATORY SYSTEM

The respiratory epithelium consists of **conducting pathways** (nasal cavities, naso- and oropharynx, larynx, trachea, bronchi, and bronchioles) and **respiratory portions** (respiratory bronchioles and alveoli). The nasal epithelium includes a region of specialized olfactory receptors. Ciliated cells appear in all portions of the respiratory system except the respiratory epithelium and move mucus and particulates toward the oropharynx. Gas exchange in the lungs takes place across a minimal barrier consisting of the capillary endothelium, a joint basal lamina, and an exceedingly thin alveolar epithelium consisting primarily of type I pneumocytes. Type II pneumocytes are responsible for the secretion of surfactant, a primarily lipid substance that facilitates respiration by reducing alveolar surface tension.

INTEGUMENTARY SYSTEM

The epidermis of thick skin consists of five layers of cells (**keratinocytes**): *stratum basale* (proliferative layer), *stratum spinosum* (characterized by tonofibrils and associated desmosomes), *stratum granulosum* (characterized by keratohyalin granules), *stratum lucidum* (a translucent layer not present in thin skin), and *stratum corneum* (characterized by dead and dying cells with compacted keratin). Specialized structures of the skin include hair follicles (found only in thin skin), nails, and sweat glands and ducts. Nonkeratinocyte epidermal cells include **melanocytes** (derived from the neural crest), **Langerhans cells** (antigen-presenting cells derived from monocytes), and **Merkel cells** (sensory mechanoreceptors). Various sensory receptors and extensive capillary networks are found in the underlying dermis.

GASTROINTESTINAL TRACT AND GLANDS

The epithelium of the gastrointestinal (GI) tract is simple and columnar throughout, except for the stratified squamous epithelia in regions of maximal friction (**esophagus** and **anus**). The **stomach** is a grinding organ with glands in the **fundus and body** that produce **mucus (surface and neck cells)**, **pepsinogen (chief cells)**, and **acid (parietal cells)**. The small intestine is an absorptive organ with folds at several levels (**plicae, villi,** and **microvilli**) that increase surface area for more efficient absorption. Cell types in the small intestine include **enterocytes (absorption)**, **Paneth cells (production of lysozyme)**, **goblet cells (mucus)**, and **enteroendocrine cells (secretion of peptide hormones)**. All of these cells originate from a single stem cell in the crypt. New cells are born in the crypt, move up the villus, and are sloughed off at the tip. The primary function of the colon, which appears histologically as crypts with prominent goblet cells and no villi, is water resorption.

The major salivary glands (**parotid, submandibular,** and **sublingual**) are exocrine glands that secrete amylase and mucus, primarily regulated by the parasympathetic nervous system. In contrast, the **pancreas** has both **exocrine (acinar cells)** and **endocrine (islet cells)** components that synthesize pancreatic juice and blood sugar–regulating hormones, respectively. The exocrine pancreas is primarily regulated by the hormones **cholecystokinin (CCK)** and **secretin**, which regulate acinar and ductal secretion, respectively.

The **liver** is also a dual-function gland whose exocrine product is **bile**, which is transported by a duct system to the **gallbladder** for storage and concentration; **bile emulsifies lipids** for more efficient enzymatic access. The endocrine products include **glucose** and major blood proteins.

ENDOCRINE GLANDS

The pituitary is formed from two embryonic sources. The **adenohypophysis** is derived from the **oral ectoderm** of Rathke's pouch and is regulated through a **hypophyseal portal system** carrying factors that stimulate or inhibit secretion. It contains **acidophils**, which produce prolactin and growth hormone (GH), and **basophils** that produce luteinizing hormone (LH), follicle-stimulating hormone (FSH), thyroid-stimulating hormone (TSH), adrenocorticotropic hormone (ACTH), and melanocyte-stimulating hormone (MSH). The **neurohypophysis** is derived from the floor of the **diencephalon** and consists of **glial cells** (**pituicytes**) and expanded terminals of nerve fibers originating in the **paraventricular** and **supraoptic nuclei** of the hypothalamus. It contains **vasopressin** and **oxytocin**.

The adrenal gland consists of two parts. The **adrenal cortex**, derived from intermediate mesoderm, and covered by a connective tissue capsule, consists of three zones: the *zona glomerulosa* produces aldosterone (a mineralocorticoid) and is regulated primarily by angiotensin II; the *zona fasciculata* and *zona reticularis* produce glucocorticoids (e.g., cortisol and weak androgens) and are regulated primarily by ACTH. The **adrenal medulla**, derived from the neural crest, synthesizes epinephrine and norepinephrine (see figure below).

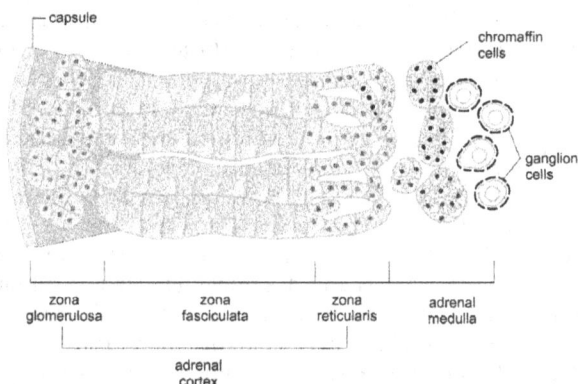

Adrenal (suprarenal) gland. The gland is covered by a connective tissue capsule and divided into a cortex containing steroid-producing cells with prominent lipid droplets (only two are drawn) and a medulla containing chromaffin cells that secrete catecholamines and neuropeptides.

The **thyroid gland** is characterized by an extracellular hormone precursor (**iodinated thyroglobulin**) in the follicles. Scattered between the follicular cells are **parafollicular cells**, which secrete **calcitonin**, a hormone that reduces blood calcium levels. The **parathyroid gland** consists primarily of **chief cells** that secrete PTH that increases blood calcium levels by stimulating osteoclastic activity and affecting kidney excretion and intestinal absorption. Endocrine cells of the **pancreatic islets** secrete primarily **insulin** and **glucagon**, hormones that regulate blood glucose. Scattered through several organ systems are enteroendocrine cells, which synthesize peptide hormones for local regulation.

URINARY SYSTEM

The filtration apparatus of the renal glomeruli consists of an expanded **basement membrane** and **slit pore** associated with podocytes. Epithelial cells of the **proximal tubule** are specialized for absorption and ion transport. They remove most of the sodium and water from urine, as well as virtually all of the amino acids, proteins, and glucose. The **brush border** of the proximal tubule cells contains proteases. The cells of the **distal tubule**, under the influence of **aldosterone**, resorb sodium and acidify the urine. Specialized cells of the distal tubule (the **macula densa**) monitor ion levels in the urine and stimulate the **juxtaglomerular cells** of the afferent arteriole to secrete **renin**, an enzyme that cleaves angiotensinogen to a precursor of angiotensin II. Collecting ducts contain light and dark (intercalated) cells; they are sensitive to **antidiuretic hormone** (**ADH**) and are the final mechanism for concentrating urine. Transitional epithelium (allowing for stretch) is found lining the calyces, renal pelvis, ureters, and urinary bladder.

MALE REPRODUCTIVE SYSTEM

The **testes** produce sperm and testosterone under the influence of **LH** and **FSH**, secreted by gonadotrophs of the anterior pituitary. The testicular epithelium contains **Sertoli cells** and precursors of sperm. **Spermatogenesis** involves the following lineage: spermatogonia (germ cells) → (**spermatocytogenesis**) → primary spermatocytes → secondary spermatocytes → (**completion of meiosis**) → **spermatids** (**spermiogenesis**) → mature sperm. Sertoli cells perform several functions: (1) maintenance of the **blood-testis barrier**, (2) phagocytosis, and (3) secretion of androgen-binding protein and inhibin, as well as Müllerian inhibiting hormone in the fetus.

The **epididymis**, like most of the male duct system, is lined by a pseudostratified epithelium characterized by modified microvilli (**stereocilia**). The **seminal vesicles** produce fructose and other molecules that activate spermatozoa.

The **prostate** is a fibromuscular organ that produces the largest fluid component of the ejaculate. Virtually all males over 70 show some form of **prostatic hypertrophy**. Prostatic malignancies are the second most common form of cancer in males.

FEMALE REPRODUCTIVE SYSTEM

The **ovaries** produce **ova**, **estrogen**, and **progesterone** under the influence of LH and FSH. Oocyte (germ cell) maturation involves several stages of follicular development (granulosa cells plus the oocyte): primordial follicle → primary follicle → secondary follicle → mature, or Graafian, follicle. In the secondary follicle, the stroma differentiates into a theca. The *theca interna* synthesizes androgens, which are converted into estradiol by granulosa cells. After ovulation, these thecal cells form the *theca lutein;* the granulosa cells become the *granulosa lutein,* which produces **progesterone** (see figure below). Human chorionic gonadotropin (hCG) in the placenta maintains the **corpus luteum** of pregnancy.

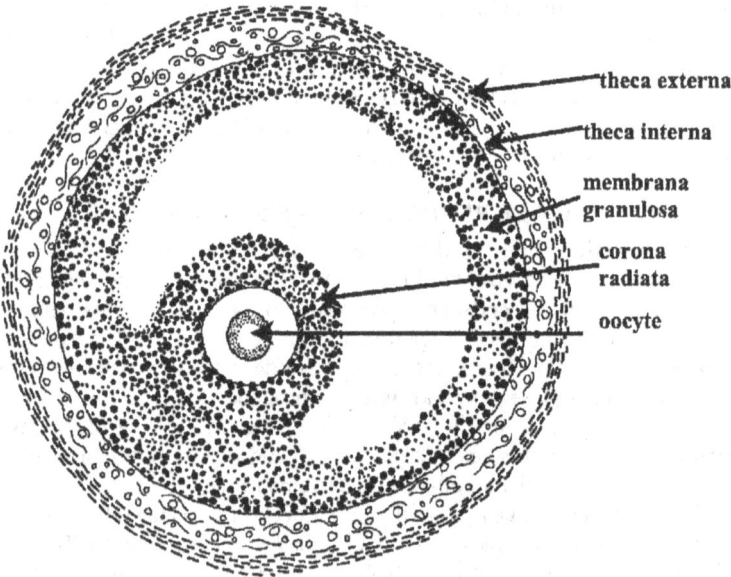

theca externa
theca interna
membrana granulosa
corona radiata
oocyte

Ovary

The **uterine endometrium** goes through a monthly cycle during which the **functionalis** is lost and replaced from the **basalis**. The **menstrual phase** occupies the first four days of the cycle (in the absence of hCG), followed by the **proliferative phase** under the influence of FSH (days 5 to 14) and then the **secretory phase** under the influence of LH (days 15 to 28). During this phase, endometrial cells accumulate glycogen preliminary to the synthesis and secretion of glycoproteins.

Vaginal epithelium is made up of stratified squamous cells and varies with maturity, phase of the menstrual cycle, pregnancy, and cancer (detected by vaginal Pap smear).

When fertilization and implantation occur, the placenta, consisting of the **chorion (fetal part)** and **decidua basalis (maternal part)**, is established for O_2/CO_2 exchange, as well as its endocrine role (e.g., conversion of androgens to estradiol, placental lactogen secretion). During parturition, oxytocin secreted by the neurohypophysis stimulates the contraction of uterine smooth muscle.

The **breast** is a resting alveolar gland except during pregnancy, when the lactiferous ducts proliferate and milk production is initiated. Milk synthesis and ejection are under the influence of prolactin and oxytocin.

EYE

The photosensitive layer of the **retina** is derived from the inner layer of the **optic cup** and contains the **rod** and **cone cells** involved in visual signal transduction. **Rhodopsin** is a visual pigment found within lamellar disks of the outer segment of the rod cell. Rhodopsin consists of **retinal** and **opsin**; photons induce an isomeric change in retinal, leading to dissociation of the retinal/opsin complex. The resulting decrease in the intracellular second messenger, guanosine 3'5'-cyclic monophosphate (**cGMP**), directs closure of membrane sodium channels and leads to **hyperpolarization** of the photoreceptor cell. This signal is transmitted to interneurons within the retina and finally to ganglion cells.

The **lens** arises from **surface ectoderm** during development. Production of lens fibers (elongated, protein-filled cells) continues throughout life without replacement. Increased opacity of the lens (**cataract**) may be caused by congenital factors, excess ultraviolet (UV) radiation, or high glucose levels.

The **choroid** and **sclera** are the supportive, protective coats of the eye. The **aqueous humor**, produced by processes of the ciliary body, flows between the lens and iris to the anterior chamber of the eye toward the iridocorneal angle, where it is drained into the **canal of Schlemm**. Blockage of the canal of Schlemm or associated structures leads to increased intraocular pressure and **glaucoma**.

EAR

The ear functions in two separate but related signal transduction systems, **audition** and **equilibrium**. The **external ear**, largely formed from the first two branchial arches, funnels sound to the tympanic membrane. The **middle ear** is made up of the **malleus, incus,** and **stapes** formed from the first two arch cartilages. The internal ear consists of a membranous and a bony labyrinth filled with **endolymph** and **perilymph,** respectively. The saccule (ventral) and utricle (dorsal), parts of the membranous labyrinth, form from the **otic vesicle** (an ectodermal invagination). The cochlea contains three spaces, the *scala vestibuli, scala media* (cochlear duct, which extends from the saccule), and *scala tympani.* The **semicircular canals** (which extend from the utricle) contain the *cristae ampulares,* made up of cupulae with hair cells embedded in a gelatinous matrix that respond to changes in direction and rate of angular acceleration. The hair cells are located within the **organ of Corti** and respond to different frequencies. In the saccule and utricle, the maculae, along with **stereocilia, kinocilia,** and **otoconia** (crystals of protein and calcium carbonate), detect changes in position with reference to gravity.

Neural Pathways

ASCENDING PATHWAYS

Afferents terminating in the spinal cord and cerebellum are generally ipsilateral. Those ascending to the thalamus and cerebral cortex terminate on the contralateral side.

The funiculi of the spinal cord are dorsal, lateral, and ventral.

Fibers within a funiculus have common origins, termination, and function.

Pain, Temperature, Tactile

Simple receptors, unmyelinated, or poorly myelinated fibers.

Enter via dorsal root and may ascend or descend a few segments.

Secondary fibers cross the midline in the ventral commissure and ascend in ventral and lateral funiculi (**ventral and lateral spinothalamic tracts**).

Terminate in **ventral posterior lateral nucleus of thalamus**.

Tertiary fibers project via the internal capsule to terminate in the postcentral gyrus.

Injury to the spinothalamic tracts results in loss of pain and temperature sensation on the opposite side of the body.

Syringomyelia interrupts pain and temperature fibers crossing in the ventral white commissure and thus results in bilateral sensory deficit.

Proprioception, Tactile Discrimination, and Stereognosis

Primary fibers arising from more complicated receptors are generally well myelinated.

Afferents enter the spinal cord via the dorsal root and ascend in the dorsal funiculus. The dorsal funiculus becomes divided into a medial **fasciculus gracilis** (sacral, lumbar, and lower thoracic inputs) and a lateral **fasciculus cuneatus** (upper thoracic and cervical inputs). Both fasciculi terminate in corresponding midbrain nuclei.

Secondary fibers cross the midline and ascend in the medial lemniscus to terminate in the **ventral posterior lateral nucleus of the thalamus**.

Tertiary fibers terminate in the postcentral gyrus.

Some primary fibers terminate in the dorsal horn. Ascending secondary fibers in the lateral funiculus form the dorsal and ventral spinocerebellar tracts that enter the cerebellum via the inferior and superior cerebellar peduncles, respectively, to terminate in the vermis.

Interruption of primary fibers in the dorsal funiculus will cause loss of proprioception, and so forth, on the same side of the body as the lesion.

Interruption of secondary fibers in the medial lemniscus will give rise to contralateral deficits.

Tabes dorsalis and pernicious anemia attack the dorsal funiculi.

Trigeminal Pathways
On reaching the brachium pontis, afferent primary trigeminal fibers divide into ascending (proprioception, two-point discrimination, light touch) and descending (pain, temperature, light touch) roots.

Primary afferents of the descending root terminate in the **sensory nucleus of CN V.**

Secondary fibers ascend through the medulla and pons as the trigeminal lemniscus to terminate in the **ventral posterior medial (VPM) nucleus** of the thalamus.

The ascending root primary tactile afferents terminate in the main sensory nucleus of CN V.

Secondary fibers ascend in the **trigeminal lemniscus** to the VPM.

Primary proprioceptive afferents from the muscles of mastication enter the pons with the motor division of the nerve and terminate in the **mesencephalic nucleus** of V.

Collaterals are given off to the motor nucleus for reflexes.

Lesion of the descending root of V and the adjacent lateral spinothalamic tract on one side of the medulla will result in pain and temperature deficits on the contralateral side of the body and the ipsilateral side of the head.

Vestibular Pathways
Primary afferents terminate in the vestibular nuclei and in the cerebellum on the same side.

Secondary fibers ascend or descend in the **medial longitudinal fasciculus** or the ventral funiculus of the spinal cord.

In the upper midbrain, fibers terminate in the motor nuclei for extraocular muscles.

In the spinal cord, secondary fibers terminate on internuncial neurons in the intermediate gray.

Unilateral lesions of the vestibular system result in movement of the head, body, and eyes (nystagmus) to the affected (ipsilateral) side. Symptoms include vertigo, nausea, and a tendency to fall to the affected side.

Visceral Afferents

Primary general visceral afferents have cell bodies in the dorsal root ganglia and terminate in the dorsal horn. Ascending secondary neurons make abundant reflex connections with autonomic and somatic pathways and terminate in the **centromedian, intralaminar,** and **parafascicular nuclei** of the thalamus.

Central processes of primary general visceral afferents associated with cranial nerves VII, IX, and X enter the solitary fasciculus and terminate in the **nucleus of the solitary tract.** Secondary fibers make reflex connections with visceral motor nuclei or ascend in the medial lemniscus to terminate in the VPM nucleus of the thalamus.

Secondary visceral tracts ascend bilaterally

DESCENDING (MOTOR) PATHWAYS

In the brain, the cell bodies of general somatic efferent neurons are located in columns ventral to the cerebral aqueduct and fourth ventricle and ventrolateral to the central canal. Special visceral efferents are located lateral and ventral to the general somatic efferents. In the spinal cord, they originate in the ventral horn.

These are lower motor neurons, or the "final common pathway." Total, or flaccid, paralysis results from destruction of peripheral nerves or motor nuclei. Destruction of upper motor neurons (from higher centers) results in spastic paralysis and hypo- or hyperreflexia.

Cerebellar Pathways

The dentate nucleus receives fibers from the Purkinje neurons of the cerebellum and projects via the superior peduncle to the reticular formation (descending limb) and to the basal ganglia/thalamus-motor cortex (ascending limb).

The cerebellum is involved with coordination of fine movements.

Lesions to the cerebellum or superior peduncle result in ataxia, hypotonia, hyporeflexia, and/or intention tremor on the same side as the lesion.

Corticospinal (Pyramidal) Pathways

Fibers arise from pyramidal neurons in layer 5 of the precentral gyrus and premotor areas and descend through the internal capsule and basis pedunculi, cross in the lower medulla, and form the **lateral corticospinal tract** in the lateral funiculus of the spinal cord. They terminate on lower motor neurons in the ventral horn or on interneurons of the central grey.

Most muscles are represented in the contralateral motor cortex. However, some (such as the muscles of the upper face and the muscles of mastication and muscles of the larynx) are represented bilaterally.

With the noted bilateral exceptions, lesion of the pyramidal tract above the decussation results in spastic paralysis, loss of fine movements, and hyperreflexia on the contralateral side.

Lesion of the corticospinal tract in the cord results in ipsilateral deficits.

The Extrapyramidal (Basal Ganglia) System

The basal ganglia (caudate, putamen, globus pallidus) and associated nuclei (e.g., substantia nigra) do not project directly to medullary or spinal lower motor neurons but to the motor cortex.

The system controls coarse, stereotyped movements. Lesions result in altered muscle tone (usually rigidity), paucity of movement, and the appearance of rhythmic tremors and writhing or jerky movements.

Reticular Pathways

Nuclei of the reticular system send ascending projections to the hypothalamus and thalamus as well as descending projections to the motor nuclei of cranial, nerves, and the intermediate gray of the spinal cord.

The reticular formation has reciprocal connections with most other areas of the CNS and produces both faciliatory and inhibitory effects on motor systems, receptors, and sensory conduction pathways.

Anatomy

UPPER EXTREMITY

- **Axillary nerve injury** results in deltoid paralysis with total inability to abduct the arm and severe impairment of flexion and extension at the glenohumeral joint.
- **Midhumeral** fracture may involve the deep brachial artery and the radial nerve as they wind about the posterior aspect of the humerus. Arterial injury produces ischemic contracture; nerve injury paralyzes the wrist extensors and extrinsic extensors of the hand.
- Except on the ulnar side, the **forearm flexor compartment** is innervated by the median nerve.
- **Scaphoid fracture** is most common because it transmits forces from the abducted hand directly to the radius. Because the blood supply enters distally, the scaphoid is especially prone to avascular necrosis.
- **Lunate dislocation** is most common in falls on the out-stretched hand, compressing the median nerve within the carpal tunnel and producing carpal tunnel syndrome.
- **Extension of the proximal phalanges** is accomplished by the extensor digitorum in the forearm, innervated by the radial nerve. **Digital extension** at the interphalangeal joints is primarily by dorsal and ventral interossei, both innervated by the ulnar nerve.
- **Proximal phalangeal flexion** is by the interossei and lumbricales (median and ulnar nerves); **middle phalangeal flexion** is by the extensor digitorum superficialis (median nerve); **distal phalangeal flexion** is by the flexor digitorum profundus (median and ulnar nerves).
- **Digital abduction** is a function of the dorsal interossei; **digital adduction** is a function of palmar interossei.
- The **ulnar artery** is the principal supply to the superficial palmar arch in the hand.
- **Lymphatic drainage** from the palmar hand and digits is toward the dorsal subcutaneous space of the hand, explaining the extreme swelling of this region that accompanies infections of the digits or volar surface.
- **Radial sensory function** is tested in the web space of the thumb; **ulnar sensory function** is tested along the fifth digit. The **digital branches** of the median and ulnar nerves lie along the sides of the fingers where they may be anesthetized (see figure below).

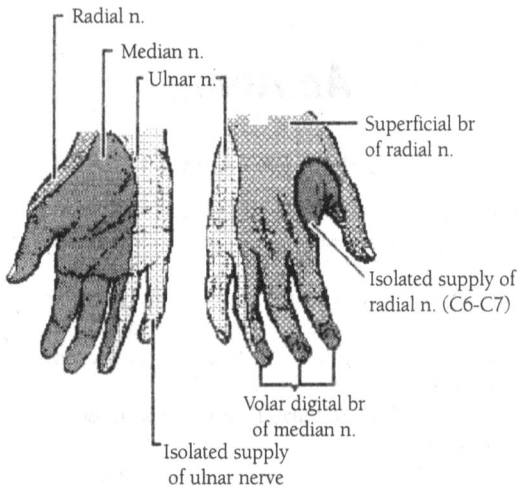

Radial n.
Median n.
Ulnar n.
Superficial br of radial n.
Isolated supply of radial n. (C6-C7)
Volar digital br of median n.
Isolated supply of ulnar nerve

Nerve Function, Tests, and Dysfunction

Nerve	Muscle Group	Reflex Test	Sign or Functional Deficit
Long thoracic	Serratus anterior		Wing scapula
Suprascapular	Supraspinatus, infraspinatus		Difficulty initiating arm abduction
Axillary	Deltoid		Inability to fully abduct arm
Radial	Extensors of forearm, wrist, proximal phalanges, and thumb	Triceps and wrist extension reflexes	Loss of arm extension; loss of forearm extension, supination, abduction; loss of wrist extension (**wrist-drop**); loss of proximal phalangeal extension and thumb extension
Musculocutaneous	Flexors of arm, forearm	Biceps reflex	Weak arm flexion, weak forearm flexion, weak forearm supination
Median	Wrist and hand flexors	Wrist flexion reflex	Paralysis of flexor, pronator, and thenar muscles; inability to fully flex the index and middle fingers (**sign of benediction**)
Ulnar	Wrist and hand flexors, phalangeal extensors		Inability to extend the distal and middle phalanges (**clawhand**); loss of thumb adduction

BACK

- **Fracture of the dens** of the axis with posterior dislocation may crush the spinal cord at the level of the first cervical vertebra with terminal paralysis of respiratory musculature.
- The **cruciform ligament** is the principal structure preventing subluxation at the atlantoaxial joint because the articular surfaces between the axis and atlas are nearly horizontal and there is no intervertebral disk.
- **Herniation** usually occurs in the fourth or fifth intervertebral disks because of the pronounced lumbar curvature and the considerable body mass superior to this region.
- The **anterior** and **posterior longitudinal ligaments** reinforce the underlying annulus fibrosus but do not meet posterolaterally, resulting in a weak area predisposed to herniation.
- **Lumbar puncture** and intrathecal anesthesia should be introduced below the third lumbar vertebra as the spinal cord usually terminates between the first and second lumbar vertebrae.
- **Posterolateral disk prolapse** impinges on the spinal nerve of the next lower vertebral level, causing symptoms associated with the dermatomic and myotomal distributions of that nerve.

Hernia Involvement, Signs, and Reflex Test

Hernia	Involvement	Signs	Reflex Test
C3-C4	C4 (Phrenic, C3-C5)	Weak diaphramatic respiration	
C4-C5	C5 (Suprascapular, C4-C6)	Weak arm abduction	
C5-C6	C6 (Musculocutaneous, C5-C6)	Weak forearm flexion	Biceps
C6-C7	C7 (Radial, C6-C8)	Weak forearm extension	Triceps
C7-C8	C8 (Ulnar, C7-T1)	Weak thumb adduction	
L1-L2	L2 (Genitofemoral, L1-L2)	Weak hip flexion	Cremaster
L2-L3	L3 (Obturator, L2-L4)	Weak hip adduction	
L3-L4	L4 (Femoral, L1-L4)	Weak leg extension	Knee jerk
L4-L5	L5 (Fibular, L4-S1)	Weak dorsiflexion	
L5-S1	S1 (Tibial, L5-S2)	Weak plantar flexion	Ankle jerk

LOWER EXTREMITY

Neurovascular Contents of the Buttock		
Quadrant	**Contents**	**Symptoms**
Upper lateral	No major vessels or nerves; A preferred location for intramuscular injection	
Upper medial	Superior gluteal neurovascular bundle	Abductor lurch
Lower lateral	Inferior gluteal neurovascular bundle	Difficulty climbing stairs or rising from a chair
Lower medial	Sciatic nerve	Foot-drop

- **Intracapsular fractures** of the femoral neck or hip dislocations tear the retinacular arteries that supply the proximal fragment; avascular necrosis may result.
- The **femoral triangle** is bounded by the inguinal ligament, the sartorius muscle, and the adductor longus muscle. A **femoral pulse** is palpable high within the femoral triangle just inferior to the inguinal ligament. The femoral vein, lying just medial to the femoral pulse, is a preferred site for insertion of venous lines.
- The **anterior cruciate ligament** is a key stabilizer of the knee joint, preventing posterior movement of the femur on the tibial plateau.
- The **medial meniscus**, being more mobile and attached to the medial collateral ligament, is most likely to be injured. Twisting movements that combine lateral displacement with lateral rotation pull the medial meniscus toward the center of the joint where it may be trapped and crushed by the medial femoral condyle.
- The **adductor canal**, the location of popliteal aneurysms, contains the femoral artery, femoral vein, and saphenous nerve.
- The **deep fibular nerve** innervates the muscles of the anterior compartment (dorsiflexors of the foot and pedal digits). The **superficial fibular nerve** innervates the lateral crural compartment (plantar flexors and everters of the foot). The **tibial nerve** innervates the posterior crural muscles which plantarflex and ·invert the foot.
- The **posterior tibial artery** descends posteriorly to the medial malleolus where the **posterior tibial pulse** is normally palpable.
- **Inversion sprains,** the most common ankle injury, involve the lateral collateral ligaments.
- The **plantar calcaneonavicular (spring) ligament** supports the head of the talus and thereby maintains the longitudinal plantar arch. Laxity of this ligament results in fallen arches or "flat feet."
- **Sensory distribution of the anterior leg:** the web space between the first and second toes is specific for the fibular nerve (L5) (see following figure).

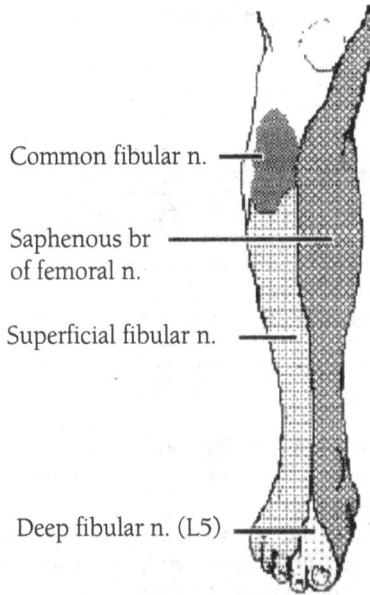

Common fibular n.

Saphenous br
of femoral n.

Superficial fibular n.

Deep fibular n. (L5)

Nerve Function, Tests, and Dysfunction			
Nerve	**Muscle Group**	**Reflex**	**Sign or Functional Deficit**
Genito-femoral	Cremaster	Cremasteric (L2-L3)	Cremaster paralysis
Femoral	Anterior thigh	Patellar (L4)	Weakness of hip flexion and loss of knee extension
Obturator	Medial thigh		Loss of thigh adduction
Superior gluteal	Gluteus medius and minimus		Abductor lurch (inability to keep pelvis level when contralateral foot is raised)
Inferior gluteal	Gluteus maximus		Difficulty rising from seated position and difficulty climbing stairs
Sciatic	Hamstrings	Hamstring (L5)	Weakness of hip extension and knee flexion
Fibular	Anterior and lateral crural compartments		Foot slap, inability to stand back on heels.
Tibial	Posterior crural compartment	Achilles (S1)	Inability to stand on tip-toes

THORAX

Thoracic Cage and Lungs

Respiratory Musculature	
Function	**Muscles**
Inspiration	External intercostals, interchondral portion of internal intercostals, and the diaphragm
Expiration	Internal intercostals proper, transverse thoracic, and abdominal muscles

- The **anterior border of the left pleural cavity** deviates laterally between the fourth and sixth ribs to form the cardiac notch—a preferred route for needle insertion into the pericardial cavity.
- When upright, excess fluid tends to collect in the **costodiaphragmatic recess.**
- Introduction of air into the pleural space results in **pneumothorax** with loss of lung ventilation. Fluid or blood produce hydrothorax and hemothorax, both of which limit expansion of the lung with reduced ventilation/perfusion ratio.
- The **right mainstem bronchus** is wider, shorter, and more vertical than the left mainstem bronchus, and therefore, is where large aspirated objects commonly lodge.
- The **right lower lobar bronchus** is most vertical, most nearly continues the direction of the trachea, and is larger in diameter than the left, and therefore, is where small aspirated objects commonly lodge, causing segmental atelectasis.
- A **bronchopulmonary segment** is defined by a segmental bronchus and accompanying segmental artery that lie centrally, as well as by intersegmental veins that form a peripheral venous plexus.
- Because the **superior segmental bronchi** of the lower lobes are the most posterior, and therefore dependent, when the patient is supine, they are most frequently involved in aspiration pneumonia (Mendelson's syndrome).

Heart
- The **transverse cardiac diameter** varies with inspiration and expiration but normally should not exceed one-half the diameter of the chest.
- An **apical pulse** is palpable at the point of maximal impulse (PMI) in the fifth intercostal space just beneath the nipple.
- **Ventricular coronary flow** occurs during ventricular diastole when a pressure differential occurs between the left ventricle and the aorta.

Cardiac Features

Landmark	Location	Contents
Coronary sulcus	Between atria and ventricles; nearly vertical behind sternum; marks the annulus fibrosus that supports the valves	Right side contains the right coronary artery, and small cardiac vein; crossed by anterior cardiac veins
		Left side contains circumflex branch of the left coronary artery and coronary sinus
Anterior interventricular sulcus	Between left and right ventricles; marks the interventricular septum	Contains the anterior interventricular branch of the left coronary artery and the great cardiac vein
Posterior interventricular sulcus	Delineates the interventricular septum, posteriorly	Contains the posterior interventricular branch of the right coronary artery and the middle cardiac vein

- The **papillary muscles** take up the slack in the chordae tendineae to maintain the competence of the valvular closure as ventricular volume is reduced during blood ejection. The valves close passively.
- A **ventricular septal defect** produces a serious right-to-left shunt with cyanosis—"blue-baby" syndrome—because left ventricular pressure exceeds that in the right ventricle. A large VSD is the principal factor in **tetralogy of Fallot.**

Heart Valves

Valve	Auscultation	Comment
Tricuspid	Right of sternum in sixth intercostal space	
Pulmonary	Left of sternum over second intercostal space	
Mitral	Apex of heart in fifth intercostal space in left midclavicular line	**Insufficiency** produces a low-pitched, late systolic blowing murmur
Aortic	Right of sternum over second intercostal space	**Stenosis** will tend to be auscultated as a high-pitched systolic murmur with possible radiation to the carotid arteries

- The **atrioventricular bundle** passes through the annulus fibrosus and descends along the posterior border of the membranous part of the interventricular septum to enter the muscular portion of the septum. It transmits electrical activity to the ventricles.

Cardiac Nodal Tissue

Node	Location	Function	Vasculature
Sinoatrial	In myocardium between crista terminalis and opening of superior vena cava	Initiates contractile event with electrical depolarization spreading throughout atrial musculature	Nodal branch of the right coronary artery
Atrioventricular	In right atrial floor near the interatrial septum	Stimulated by atrial depolarization; it leads into the atrioventricular (A-V) bundle to synchronize ventricular depolarization	Branch of right coronary artery near the posterior inter-ventricular branch

- **Autonomic pathways** consist of two motor neurons, a myelinated preganglionic (presynaptic) neuron and an unmyelinated postganglionic (postsynaptic) neuron.

Summary of Autonomic Pathways

Division	Presynaptic Pathway	Postsynaptic Pathway	Effect
Sympathetic	From spinal levels T1-L2 along the ventral root; Reach the chain of sympathetic ganglia via white rami communicantes	1. Fibers that synapse return to the spinal nerve via a gray ramus to mediate cutaneous piloerection, vasoconstriction, and sudomotor activity 2. Fibers that do not synapse pass through the chain as splanchnic nerves to synapse in prevertebral ganglia; from these ganglia, postsynaptic neurons run in perivascular plexuses to innervate visceral target tissues	**Adrenergic neurotransmission** increases heart rate, increases stroke volume, dilates coronary and pulmonary arteries

Division	Presynaptic Pathway	Postsynaptic Pathway	Effect
Para-sympathetic	Presynaptic cell bodies are located in the dorsal vagal nuclei of the brain; The myelinated synaptic axons form cranial nerve X, the vagus nerve	Postganglionic cell bodies lie in numerous ganglia close to the target organ	**Cholinergic neurotransmission** decreases heart rate, decreases stroke volume, and produces bronchoconstriction

Pain Referral from Thoracic Viscera		
Organ	Referral Area	Pathway
Pericardial cavity	T1-T5: upper and midthorax	Intercostal nerves T1-T5
Heart	T1-T4: upper thorax, postaxial brachium	Cervical and thoracic splanchnic nerves
Thoracic esophagus	T1-T5: thorax and epigastric region	Thoracic splanchnic nerves
Diaphragm		
Central	C3-C5: neck and shoulder	Phrenic nerve
Marginal	T5-T10: thorax	Intercostal nerves

ABDOMEN

Abdominal Wall

Dermatomal Landmarks	
Dermatome	Region
T4	Nipple
T10	Umbilicus
T12	Pubis

- The **abdominal musculature** has three distinct layers that take three different directions. The external oblique muscle, internal oblique muscle, and transverse abdominis muscle may be sequentially split and retracted so that extensive suturing is unnecessary to provide a strong repair (McBurney's incision).

- Because the **linea alba** is relatively avascular, incisions may not heal well and predispose to epigastric herniation.
- **Above the arcuate line,** the anterior leaf of the **rectus sheath** is formed by fusion of the external oblique and internal oblique aponeuroses; the posterior leaf is formed by fusion of the internal oblique and transverse abdominis aponeuroses.
- **Below the arcuate line,** the anterior leaf of the rectus sheath is formed by fusion of all three aponeuroses and there is no posterior leaf.
- The **inferior epigastric artery** passes into the rectus sheath at the arcuate line. This is a potential site for spigelian herniation into the rectus sheath.

Hernia Characteristics

Hernia	Pathway
Direct inguinal	Through the inguinal triangle bounded by inguinal ligament, inferior epigastric artery, and rectus abdominis—therefore, medial to the inferior epigastric artery. Exits through the superficial inguinal ring *adjacent* to the spermatic cord. Usually acquired
Indirect inguinal	Through the deep inguinal ring and along the inguinal canal— therefore, lateral to the inferior epigastric artery. Exits through the superficial ring *within* the spermatic cord. Usually congenital
Femoral	Passes inferior to the inguinal ligament through the femoral ring into the thigh. More prevalent in women

GI Tract

Characterization of Abdominal Structures by Location and Support

Characterization	Organ
Peritoneal (supported by mesentery)	Abdominal esophagus, stomach, superior duodenum, liver, pancreatic tail, jejunum, ileum, a variable portion of the cecum, appendix, transverse colon, and sigmoid colon
Secondarily retroperitoneal (adherent)	Descending and inferior duodenum, pancreatic head and body, ascending colon, and descending colon. These may be surgically mobilized with an intact blood supply
Extraperitoneal	Thoracic esophagus, rectum, kidneys, ureters, and adrenal glands

- **Peptic ulceration** of the lower esophagus, stomach, or superior duodenum is referred along the greater splanchnic nerve to the fifth and sixth dermatomes which include the epigastric region.
- The **hepatic triangle**, bounded by the cystic duct, gallbladder, and common hepatic duct, contains the cystic arteries and right hepatic artery with potential for extensive variation.
- The **duodenal papilla** usually contains the hepatopancreatic ampulla, formed by the joining of the common bile duct and the pancreatic duct. If blocked by a stone, pancreatitis may develop.
- The **tail of the pancreas** contains most of the pancreatic islets (of Langerhans), a consideration in pancreatic resection.
- **Ileal (Meckel's) diverticulum** is found in about 3% of the population, located within 3 ft of the ileocecal junction (on the antimesenteric side of the ileum), and usually less than 3 in. long. Peptic ulceration of adjacent ileal mucosa and volvulus are complications.
- The **hepatic portal vein** directs venous return from the gastrointestinal tract to the liver.
- Because the **hepatic portal system** has no valves, blood need not flow toward the liver. Liver disease (such as cirrhosis) or compression of a vein (as in pregnancy or constipation) results in blood shunting through the anastomotic connections to the systemic venous system.

Portal-Systemic Anastomoses Occur in Several Areas		
Location	**Anastomotic Connections**	**Signs & Symptoms**
Esophagus	Azygos veins with left gastric and short gastric veins	Esophageal varices, intractable hematemesis
Umbilicus	Paraumbilical veins with superior and inferior epigastric veins	Caput medusae
Rectum	Superior rectal vein with middle and inferior rectal veins	Internal and external hemorrhoids

Kidneys, Ureters, and Adrenal Glands

- The **renal fascia** (the false capsule or Gerota's fascia) is a discrete fascial layer that surrounds each kidney. Paranephric fat outside this capsule and perinephric fat inside this fascial layer support the kidney.
- **Minor calyces** receive one or two pyramids before fusing into major calyces. Two to four minor calyces join to form **major calyces** that coalesce to form the renal pelvis.

- The **ureters** narrow at three points—at the renal pelvis, at the pelvic brim, and at the bladder. Kidney stones may lodge at these locations with pain referred, respectively, to the subcostal, inguinal, and perineal regions.
- **Adrenal arteries** arise from the inferior phrenic arteries, the aorta, and the renal arteries. The right adrenal vein usually drains medially into the inferior vena cava; the left adrenal vein usually drains inferiorly into the left renal vein.
- The **superior lumbar trigone** (a posterior approach to the kidneys, suprarenal glands, and upper ureters) is bounded by the quadratus lumborum muscle, superior border of the internal oblique muscle, and the twelfth rib.

Pain Referral from Abdominal Viscera

Organ	Referral Area	Pathway
Diaphragm		
Central	C3-C5: neck and shoulder	Phrenic nerve
Marginal	T5-T10: thorax	Intercostal nerves
Stomach, gallbladder, liver, bile duct, superior duodenum	T5-T9: lower thorax, epigastric region	Celiac plexus to greater splanchnic nerve
Inferior duodenum, jejunum, ileum, appendix, ascending colon, transverse colon	T10-T11: umbilical region	Superior mesenteric plexus to lesser splanchnic nerve
Kidneys, upper ureters, gonads	T12-L1: lumbar and ipsilateral inguinal regions	Aorticorenal plexus to least splanchnic nerve
Descending colon, sigmoid colon, mid ureters	L1-L2: pubic and inguinal regions, anterior scrotum or labia, anterior thigh	Aortic plexus to lumbar splanchnic nerves

PELVIS

Perineum

- The **external anal sphincter,** innervated by the pudendal nerve, provides the brief voluntary contraction necessary to counter the passage of a peristaltic wave.
- The **rectal submucosal venous plexus** forms anastomotic connections between the middle rectal veins that drain directly into the internal iliac veins and the superior rectal veins that drain into the hepatic portal system. This is a site for varices (hemorrhoids).
- The **internal pudendal arteries** are the sole supply of both male and female erectile tissue.

- The **deep dorsal vein** provides venous return from the penis or clitoris by passing through the urogenital diaphragm and draining into the prostatic or vesicle venous plexus, respectively.
- The **cremaster muscle** of the spermatic cord is innervated by the genital branch of the genitofemoral nerve. This provides the efferent limb for the cremaster reflex (L1-L2), the elevation of the testes within the scrotum when the inner thigh is scratched.
- The **cavity of tunica vaginalis** is a potential space that represents the detached portion of the peritoneal cavity that surrounds the testis except at the mesorchium.
- Because the superficial perineal space is limited by fascial attachment to the deep transverse perineal muscle, extravasations of blood or urine will not pass into the anal triangle.

Contents of the Perineal Spaces are Gender Specific

Gender	Superficial Perineal Space	Deep Perineal Space
Male	Testes, crura of penis, bulb of penis, penile urethra, superficial transverse perineal muscles	Deep transverse perineal external urethral sphincter, bulbourethral glands, membranous urethra
Female	Crura of the clitoris, vestibular bulbs, superficial transverse perineal muscles, greater vestibular glands	Deep transverse perineal muscle, external urethral sphincter, urethra

- The **male external urethral sphincter** is formed by the deep transverse perineal muscle completely surrounding the membranous urethra. The **female external urethral sphincter** is formed by muscle fascicles of the deep transverse perineal muscle that arch anterior to the urethra but do not pass posterior because the urethra is embedded in the adventitia of the anterior vaginal wall. The arrangement in the female perineum predisposes to urinary stress incontinence.
- A **pudendal block** can be effected by injecting an anesthetic into the vicinity of the pudendal nerve in the pudendal canal close to the ischial spine.

Pelvic Autonomic Function

Function	Sympathetic	Parasympathetic
Emission	L1-L2: lumbar splanchnic nerves,	
Erection	hypogastric plexus, pelvic plexus, cavernous plexus	
Ejaculation		S3-S5: pelvic splanchnic nerves

Pelvic viscera

- The **female pelvis** is less massive, the subpubic angle is greater (almost 90°), and the pelvic inlet more ovoid than the male pelvis.
- The **obstetric conjugate** is the least anteroposterior diameter of the pelvic inlet from the sacral promontory to a point a few millimeters below the superior margin of the pubic symphysis.
- The **transverse midplane diameter,** measured between the ischial spines, is the smallest dimension of the pelvic outlet.
- The **levator ani muscle** forms most of the pelvic floor and its puborectalis portion (rectal sling) is the principal mechanism for maintenance of fecal continence when the rectum is full.

Characterization of Pelvic Structures by Location and Support	
Characterization	**Organ**
Peritoneal (supported by mesentery)	Sigmoid colon, uterus, uterine tubes, ovaries, testes
Extraperitoneal	Rectum, anal canal, urinary bladder, cervix, prostate gland, seminal vesicles

- The **rectum** is usually empty because feces are stored in the sigmoid colon. Movement of feces into the rectal ampulla generates the urge to defecate.
- **Metastatic carcinoma of the rectum** may be widely disseminated within the abdomen, pelvis, and inguinal region. The upper rectum drains along the superior rectal lymphatics. The midrectum drains along the middle rectal lymphatics. The lower rectum drains along the inferior rectal lymphatics and then along both internal and external pudendal lymphatic channels.

Urinary Bladder Innervation is by Both Sympathetic and Parasympathetic Routes	
Function	**Pathway**
Sensory awareness of bladder fullness	Hypogastric nerve (sympathetic pathways) to spinal segments T12-L2
Afferent limb of the detrusor (bladder-emptying) reflex	Pelvic plexus and pelvic splanchnic nerves (parasympathetic pathways) to spinal segments S2-S4
Efferent limb of the detrusor reflex	Pelvic splanchnic nerves (parasympathetic pathways) from S3-S5

- **Urinary continence** of the partially full to full urinary bladder is a function of the external urethral sphincter.
- A **patent urachus** (rare) allows reflux of urine through the umbilicus.
- The **testes** develop as retroperitoneal structures, but become peritoneal (supported by mesorchium) in the scrotum. A long **mesorchium** may predispose to testicular torsion with high potential for testicular ischemia and necrosis.
- The testicular **pampiniform plexus** functions as a countercurrent heat exchanger that maintains testicular temperature a few degrees below core body temperature.
- **Compression of the left testicular vein** by a full sigmoid colon produces varices of the pampiniform plexus on the left side; fertility may diminish.
- **In the male, palpable per rectum** are posterior and lateral lobes of the prostate gland, seminal vesicles if enlarged, and bladder when filling.
- Each **uterine artery** crosses immediately superior to a ureter in the transverse cervical ligament—an important surgical consideration.
- **Normal uterine position** is anteflexed (uterus bent forward on itself at the level of the internal os) and anteverted (angled approximately 90° anterior to the vagina), lying on the urinary bladder.
- **In the female, palpable per vagina** are the cervix and ostium of the uterus, the vagina, the body of the uterus if retroverted, the rectouterine fossa, and variably the ovary and uterine tubes.
- The **lymphatic drainage from the vagina** is by three routes: the external and internal iliac nodes from the upper third of the vagina; the internal iliac nodes from the middle third of the vagina; and the internal iliac nodes as well as the superficial inguinal nodes from the lowest third.

Pain Referral from Pelvic Viscera

Organ	Referral Area	Pathway
Testes and ovaries	T10-T12: umbilical and pubic regions	Gonadal nerves to aortic plexus and then to lesser and least splanchnic nerves
Middle ureters, urinary bladder, uterine body, uterine tubes	L1-L2: pubic and inguinal regions, anterior scrotum or labia, anterior thigh	Hypogastric plexus to aortic plexus and then to lumbar splanchnic nerves
Rectum, superior anal canal, pelvic ureters, cervix, epididymis, vas deferens, seminal vesicles, prostate gland	S3-S5: perineum and posterior thigh	Pelvic plexus to pelvic splanchnic nerves

HEAD AND NECK

Somatic Portions

- **The scalp layer of loose connective tissue** between the epicranial aponeurosis and the periosteum forms the subaponeurotic or "danger" space. Emissary veins connect with the dural sinuses with potential for hematogenous spread of infection through the calvaria.
- **Cranial fractures** preferentially pass through cranial foramina injuring the contained nerves.

Principal Foramina of the Anterior Cranial Fossa

Foramen	Contents	Result of injury
Olfactory	Olfactory nerves	Anosmia
Foramen cecum	An emissary vein	

Principal Foramina of the Middle Cranial Fossa

Foramen	Contents	Result of injury
Optic canal	CN II	Unilateral blindness
	Ophthalmic artery	Ischemic unilateral blindness
Superior orbital fissure	CN III	Ophthalmoplegia
	CN IV	Inability to look down and out
	CN V$_1$	Unilateral loss of blink reflex
	CN VI	Inability to abduct eye
	Superior ophthalmic vein	Retinal engorgement
Foramen rotundum	CN V$_2$	Loss of sneeze reflex
Foramen ovale	CN V$_3$	Masticatory paralysis, loss of jaw-jerk reflex
Foramen spinosum	Middle meningeal artery	
Foramen lacerum	Nothing (except occasionally the greater superficial petrosal nerve)	
Hiatus of the facial canal	Gr. superficial petrosal n.	Dry eye, loss of submandibular and sublingual secretion

Principal Foramina of the Posterior Cranial Fossa

Foramen	Contents	Result of Injury
Internal auditory meatus	CN VII	Facial paralysis
	CN VIII	Auditory and vestibular deficits
Jugular foramen	CN IX	Loss of gag and carotid reflexes
	CN X	Loss of cough reflex; paralysis of laryngeal muscles and some palatine muscles
	CN XI	Inability to shrug shoulders
	Internal jugular vein	
Anterior condylar canal	CN XII	Paralysis of tongue muscles; lingual deviation toward side of injury upon protrusion

CSF Is Produced by the Choroid Plexuses that Project into the Ventricles of the Brain

CSF Production	Through	Into
Lateral ventricles	Foramina of Monro	Third ventricle
Third ventricle	Iter (cerebral aqueduct)	Fourth ventricle
Fourth ventricle	Foramina of Magendie and Luschka	Cisterna magna of subarachnoid space

From	Through	CSF Uptake
Subarachnoid space	Arachnoid villi	Superior sagittal venous sinus

- The **cerebral aqueduct** is prone to occlusion, leading to hydrocephalus.

Cranial and Cerebral Hematomas

Hematoma	Prognosis	Location	Cause
Epicranial	Resolves	Subaponeurotic space	Superficial vessels
Epidural	Life-threatening	Epidural space	Torn middle meningeal artery
Subdural	Less serious	Subdural space	Torn cerebral vein
Subarachnoid	Lethal	Subarachnoid space	Torn cerebral artery, cerebral aneurysm
Subpial	Usually resolves	Cerebrum	Cerebral contusion

- **Regions of the orbit** that are prone to fracture include the ethmoid lamina papyracea and the maxilla about the infraorbital groove.
- Contraction of the **orbicularis oculi muscle**, innervated by the facial nerve, produces the blink.

Orbital Muscle Function and Innervation

Muscle	Primary Function	Secondary Functions (normally balance)	Innervation
Pupil	Constriction		CN III parasympathetic
	Dilation		Sympathetic chain
Ciliary body	Accommodation		CN III parasympathetic
Superior tarsal muscle	Augment levator palpebrae superioris		Sympathetic chain
Levator palpebrae superioris	Elevate eyelid		CN III (Oculomotor)
Medial rectus	Adduction		CN III (Oculomotor)
Superior rectus	Elevation	Adduction, intorsion	CN III (Oculomotor)
Inferior oblique	Elevation	Abduction, extorsion	CN III (Oculomotor)
Inferior rectus	Depression	Adduction, extorsion	CN III (Oculomotor)
Superior oblique	Depression	Abduction, intorsion	CN IV (Trochlear)
Lateral rectus	Abduction		CN VI (Abducens)

- **Parasympathetic innervation** to the pupil originates in the Edinger-Westphal nucleus and travels with the oculomotor nerve. Temporal lobe herniation (from tumor, hematoma, or edema) compresses the oculomotor nerve within the tentorial notch, causing a dilated pupil that is unresponsive to light.

Special Sensory Tests and Dysfunction

Nerve	Foramen	Dysfunction	Test
CN I (Olfactory)	Cribriform plate	Anosmia	Whiff of clove
CN II (Optic)	Optic canal	Blindness	Optic field tests
CN VIII			
Cochlear	Internal auditory meatus	Deafness	Hearing threshold
Vestibular	Internal auditory meatus	Balance	Nystagmus

- Paralysis of the **stapedius muscle,** as a result of facial nerve palsy, produces hyperacusis.

Visceral Portions

- The **infrahyoid muscles,** innervated by the ansa cervicalis (C1-C3), stabilize the hyoid bone and larynx during deglutition and phonation.
- The **pretracheal space,** deep to the pretracheal fascia, surrounds the trachea and thyroid gland, but is anterior to the esophagus. Infection in this space may track into the superior mediastinum.
- The **retropharyngeal (retrovisceral) space** lies posterior to the oropharynx and esophagus and is defined by septa from the pretracheal fascia. Infection within this space may track into the posterior mediastinum.
- The **mandibular neurovascular bundle** enters the mandibular foramen adjacent to the lingula, the point of minimal movement. It may be anesthetized by directing a needle posteriorly through the buccal wall just lateral to the pterygomandibular raphe.
- The **deep cervical nodes** receive lymph from the anteroinferior portion of the face, the nasal cavities, and the oral cavity.
- The **nasal vestibule** (the most common site for nosebleeds) receives vascular branches from internal and external carotid arteries.
- The **palatine tonsil** receives vascular branches from the maxillary, facial, and lingual arteries.
- **Abduction of the vocal cords** is a function of the posterior cricoarytenoid muscle only, innervated by the recurrent laryngeal nerve.

Branchiomeric Nerve Functions and Tests				
Nerve	**Course**	**Sensory**	**Motor**	**Test**
CN V (trigeminal)				
V1	Superior orbital fissure, supraorbital notch	Forehead	None	Blink reflex
V2	Foramen rotundum, maxillary foramen	Midface	None	Sneeze reflex

Continued

Nerve	Course	Sensory	Motor	Test
V3	Foramen ovale, mandibular foramen, mental foramen	Anterior pinna, jaw	Muscles of mastication, mylohyoid ant. belly of digastric, tensor palatini and tensor tympani	Jaw jerk
CN VII (facial)	Internal auditory meatus, facial canal, stylomastoid foramen	Concha of ear, taste anterior ⅔ of tongue via chorda tympani	Muscles of facial expression, stylohyoid, post. belly of digastric, tensor tympani, parasympathetic to lacrimal, nasal, palatine, lingual and submandibular glands via gr. superficial petrosal nerve	Blink reflex
CN IX (glosso-pharyngeal)	Jugular foramen	External auditory, meatus, oropharynx, carotid body and sinus, taste posterior ⅓ of tongue	Stylopharyngeus muscle, parasym-pathetic to parotid gland via tympanic and lesser super-ficial petrosal nerves	Gag reflex, Carotid reflex
CN X (vagus)	Jugular foramen	External auditory meatus, larynx, taste from epiglottis, aortic body	Palatine muscles, pharyngeal muscles, laryngeal muscles	Phonation

Nerve Functions and Tests

Nerve	Foramen	Sensory	Motor	Test
CN XI (spinal accessory)	Foramen magnum, jugular foramen	None	Sternomastoid Upper trapezius	Turn head to opposite side
CN XII (hypoglossal)	Hypoglossal canal	None	Intrinsic and extrinsic tongue muscles	Protrudes straight

• **Sensory innervation of the face** is by the trigeminal nerve (see figure below).

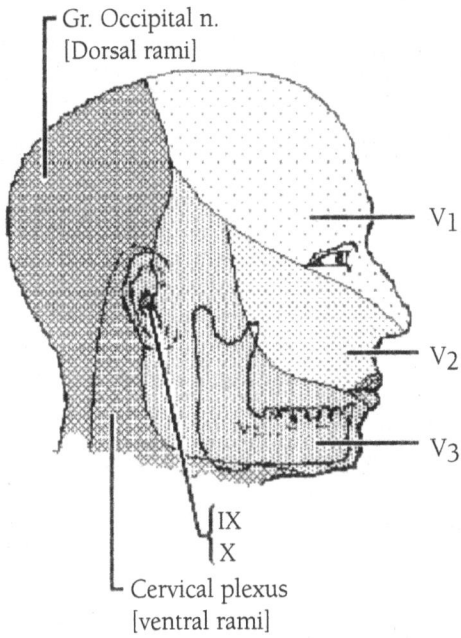

High-Yield Facts in Behavioral Science

Anxiety Disorders: Among the most common psychiatric disorders, affecting upward of 15% of the population:
- Phobic disorders affect 8 to 10%
- Generalized anxiety disorder found in 5%
- Panic disorder found in 1 to 3%
- Obsessive-compulsive disorder found in 1 to 3%

Attention Deficit Hyperactivity Disorder (ADHD): Inability to maintain attention, poor impulse control, low frustration tolerance, excessive activity; characterized by the following:
- Occurs in 5 to 12% of school age boys
- M:F = 6:1
- 50% have learning disorders
- Extends into adulthood in 33% of cases
- Twin studies show strong genetic influence
- Rx: behavior therapy and/or methylphenidate or dextroamphetamine

Bipolar Affective Disorder
- Lifetime prevalence rate of 0.6 to 1.1%
- Equally distributed among males and females
- Two subtypes, Bipolar I (recurrent major depressive episodes with manic episodes) and Bipolar II (recurrent major depressive episodes with at least one hypomanic episode)
- Characteristic symptoms are labile affect with predominant euphoria, expansiveness, grandiosity, racing thoughts, flight of ideas, mood-congruent delusions, loosening of associations, increased motor activity and rate of speech, increased spending, pressured speech

Cardiovascular Disease: Reduced morbidity/mortality rates over the past 25 years:
- 40% of reduction is the result of medical developments (drugs, technology, new Dx and Rx, surgery, transplants, training)

- 50% of reduction is attributed to awareness of risk factors, individual behavior, and lifestyle changes to reduce risks
- 30% is attributed to exercise, less fat intake, lower cholesterol, and better diet
- 24% to reduced smoking

Classical Conditioning: Pairing a natural reaction to an internal or external stimulus with a specific behavior (Pavlov).

Compulsion: Repetitive behaviors or thoughts that patient feels driven to perform to produce temporary relief from anxiety (handwashing, counting behaviors, checking and rechecking).

DSM-IV: Biopsychosocial diagnostic symptoms for psychiatric/mental health disorders:

Axis I: Majority of clinical psychiatric and psychological disorders; schizophrenia, mood disorders, anxiety, physical and substance abuse, noncompliance

Axis II: Personality disorders and mental retardation; schizotypal, antisocial, and developmental disorders, such as mental retardation, autism, dyslexia

Axis III: General medical disorders; arteriosclerotic heart disease, psoriasis

Axis IV: Psychosocial and environmental stressors before and after mental disorder (severity is quantified on an ordinal scale)

Axis V: Patient's overall global assessment of functioning, prior year and present

Fetal Risk from Alcohol: Risks from alcohol consumption during pregnancy: pre- and postnatal development, retardation, microcephaly, facial abnormalities, limb dislocation, heart and lung fistulas.

Grief: Typically lasts 6 to 12 months, characterized by
- Shock (I can't believe it)
- Denial (can't be; it didn't happen)
- Guilt (why didn't I do/say . . . ; if only I had . . .)
- Acceptance (I'm doing this for them; my life will be better).

Heritability: In older twins, highest degree of heritability is general cognitive ability (80%), followed by interpersonal skills and domestic skills.

Immune System: Function can be altered by behavioral factors (e.g., stress, depression, isolation, bereavement, diet, suppressed emotions, relaxation, conditioning, anxiety, anger).

Incidence: The number of new cases in a population over a period of time.

Infant Deprivation of Affection and Contact: Results in poor socialization and language skills, decreased muscle tone, anaclitic depression, lack of inquisitive investigation, and increased risk of infection.

Infant Mortality Rate: In 1995, the U.S. infant mortality rate
- Was 7.6 deaths per 1000 live births
- Ranked twenty-fifth among industrialized nations
- Was 2.4 times higher for black infants than for white infants

Kübler-Ross Stages of Dying: Stages progress at different rates and often vacillate:
- Denial (can't be; not me)
- Anger (why me; not fair)
- Bargaining (please, God, I promise I'll never do it again; I'll dedicate my life to . . . if I can survive this)
- Grief (I feel so sad/down/depressed)
- Acceptance (I can take it, I've had a good life; I must plan for those who live on).

Major Depressive Disorder
- Lifetime prevalence rate is 3 to 6%
- Two to 3 times more common in adolescent and adult females than males
- Major signs and symptoms include depressed mood, anhedonia, appetite change, sleep disturbance, psychomotor retardation, loss of energy, feelings of worthlessness and guilt, decreased concentration, suicidal ideation

Meta-analysis: Pooling data from several studies investigating the same hypothesis to achieve greater statistical power (not able to overcome methodological limitations or bias of an individual study).

Motor Vehicle and Firearm Deaths: The two leading causes of injury death in the U.S. in 1995; 29 and 24% of all injury deaths, respectively. Poisoning was the third leading cause of injury death at 11%. In 1994, the firearm injury death rate among U.S. males 15 to 24 years of age was 32% higher than the motor vehicle injury death rate.

Nursing Home Residents: Between 1992 and 1995
- The number of nursing home residents per 1000 elderly population declined 8% to 408
- Nursing home occupancy rates declined 5 to 81%

Obsession: Having recurrent, intrusive, and persistent thoughts, impulses, or images that cannot be ignored by logical effort; associated with anxiety, core fear (germs, death, insecurity).

Obsession and compulsion: Both usually are concealed and associated with anxiety, impairment of daily functions and time-consuming behavior.

Operant Conditioning: Reinforcing desired target behavior that may not be natural to the individual; antecedents and consequences determine behavior (Skinner).

Overweight Americans: The percentage of overweight
- Adolescents 12 to 17 years of age increased from 6% in the years between 1976 and 1980 to 12% between 1988 and 1994
- Children 6 to 11 years old increased from 8 to 14%
- Adults increased from 25 to 35%

P < 0.05: Probability is less than 5% that results have occurred by chance.

Poverty: In 1995, 36.4 million people lived in poverty in the U.S.; 40% of these were children.

Poverty in Women: Women constitute a majority of the poor in all societies. In the U.S.,
- One-third of families headed by women live in poverty
- One-half of African American and Latino women live in poverty
- 20% of women over 65 years of age live in poverty

Prevalence: The total number of cases that can be found in a population.

Primary Prevention: Prevents the occurrence of disease (e.g., vaccination).

Reliability: A test or measurement produces the same result or score if remeasured.

Responses to Stress: Physiologic responses to stress include sympathetic activation, suppression of immune system, decreased NK cell activity, decreased B cells and cytotoxic T lymphocytes, disruption of DNA repair, suppression of interferon, release of epinephrine, norepinephrine, and enkephalins, activation of latent viruses, cardiac arrhythmia, angina pectoris, increased cortisol and circulating neutrophils, increased use of drugs, smoking, and alcohol, increased heart rate, blood pressure, and respiration.

Schizophrenia
- Incidence of 1% throughout all societies in the world
- Three syndromes comprise positive, negative, and disorganization symptoms
- Characteristic symptoms are delusions, hallucinations, disorganized speech and thought processes, grossly disorganized or catatonic behavior, affective flattening, alogia, avolition

Schizophrenia Heritability: One percent in population; first-degree relatives, 10% (whether reared together or apart); fraternal twins, 17%; identical twins, 48%; may be linked to chromosome 6.

Secondary Prevention: Early detection of disease (e.g., Pap smear, mammogram).

Stages of Development: Typical stages of lifetime psychological development; tasks versus consequences of not dealing with them successfully (Erik Erikson):

Stage	Task versus Consequence
0–1 yr:	Trust versus mistrust
1–3 yrs:	Autonomy versus shame/doubt
3–6 yrs:	Initiative versus guilt
6–12 yrs:	Industry versus inferiority
12–20 yrs:	Identity versus role confusion
20–30 yrs:	Intimacy versus isolation
30–65 yrs:	Generativity versus self-absorption
65+ yrs:	Integrity versus despair

Suicide: Major risk factors include white, male, no spouse, prior attempts, presence of plan, alcohol or drug use, family history, medical illness, taking three or more prescription drugs.

Support Groups with Psychotherapy: Women being medically treated for active breast cancer participating in a support group with psychotherapy
- Extend their survival time by approximately 18 months
- Have double the survival time of women with medical therapy but with no support group

Symptoms of Heroin Addiction: Abstinence syndrome, including dilated pupils, lacrimation, sweating, irritability, rhinorrhea, muscle aches, needle-stick scars; often leading to hepatitis, overdose, abscesses, hemorrhoids, HIV, and right-sided endocarditis.

Tertiary Prevention: Reduces disability from disease (e.g., insulin, rehabilitation).

Type A Behavior: Most significant factors are anger and hostility.

Type I Error: Concluding that a difference exists when it does not.

Type II Error: Concluding that a difference does not exist when it does.

Unintentional Injury: In 1995
- Unintentional injury accounted for 61% of all injury deaths; suicide for 21%; homicide for 15%
- Unintentional injury mortality and suicide rates were highest for the elderly
- Homicide rates were highest for adults 20 to 24 years of age

U.S. Elderly Population: By the year 2000 with U.S. population of 300 million, 35 million will be over age 65 (13% increase over 1995), with the greatest increase over age 85 (expect similar increase in health care costs).

Validity: A test or measurement truly measures what it is intended to measure.

High-Yield Facts in Biochemistry & Genetics

HORMONAL CONTROL OF METABOLISM

Metabolism is precisely regulated by hormones controlling the level of blood fuels and their delivery to tissues. The primary control hormones of metabolism are insulin and glucagon. Epinephrine has effects similar to those of glucagon, except that glucagon has a greater effect on the liver while epinephrine has a greater effect on muscle. Blood levels of glucose, amino acids, fatty acids, and ketone bodies are maintained by variations in the [insulin]/[glucagon] ratio. When blood sugar is high, the ratio increases and insulin signals the fed state, promoting anabolic activities. The ratio decreases as glucagon is released to direct catabolic activities when blood glucose falls between meals, during fasting, and during starvation. Epinephrine or norepinephrine is released during exercise to promote catabolism of glucose and fat that supports muscular activity. Under normal conditions, the very precise interplay between insulin and glucagon maintains homeostatic blood fuel levels at about: glucose, 4.5 mM; fatty acids, 0.5 mM; amino acids, 4.5 mM; ketone bodies, 0.02 mM. Blood levels of ketone bodies and fatty acids rise during fasting or during starvation, with blood glucose levels being maintained. However, during uncontrolled juvenile diabetes, blood glucose levels rise greatly. The lack of insulin in this disease otherwise mimics starvation. The activity of various pathways during different metabolic states is summarized in the following table.

ACTIVITY OF METABOLIC PATHWAYS

Pathway	Fed	Fasted	Diabetes
Glycogen synthesis	+	–	–
Glycolysis (liver)	+	–	–
Triacylglyceride synthesis	+	–	–
Fatty acid synthesis	+	–	–
Protein synthesis	+	–	–
Cholesterol synthesis	+	–	–
Glycogenolysis	–	+	+
Gluconeogenesis (liver)	–	+	+
Lipolysis	–	+	+
Fatty acid oxidation	–	+	+
Protein breakdown	–	+/–	+/–
Ketogenesis (liver)	–	+	+
Ketone body utilization (non-hepatic tissues)	–	+	+

KEY FACTS ABOUT INHERITANCE

- Human gametes have 23 chromosomes (haploid chromosome number $n = 23$), while most somatic cells have 46 chromosomes (diploid chromosome number $2n = 46$).

- Genes occupy sites on chromosomes (loci) and occur in alternative forms (alleles).

- Mendelian diseases exhibit autosomal dominant, autosomal recessive, or X-linked inheritance, while multifactorial diseases (e.g., cleft palate, diabetes mellitus, schizophrenia, hypertension) are determined by multiple genes plus the environment.

- Characteristics of autosomal dominant diseases include a vertical pedigree pattern, affliction of both males and females, variable expressivity (variable severity among affected individuals), frequent new mutations, and a 50% recurrence risk for offspring of affected individuals (see pedigree A on chart). *Corollary:* germ-line mosaicism may produce affected siblings with autosomal dominant disease when neither parent is affected.

- Characteristics of autosomal recessive diseases include a horizontal pedigree pattern, affliction of males and females, frequent consanguinity (inbreeding), frequent carriers (heterozygotes without manifestations of disease), and a 25% recurrence risk for carrier parents (see pedigree B on chart). *Corollary:* normal siblings of individuals with autosomal recessive disease have a 2/3 chance of being carriers.

• Characteristics of X-linked recessive diseases include an oblique pedigree pattern, affliction of males only, frequent female carriers, and a 25% recurrence risk for carrier females (see pedigree C on chart). *Corollary:* Haldane's law predicts a 2/3 chance that the mother of an affected male with X-linked recessive disease is a carrier (and a 1/3 chance the affected male represents a new mutation).

• Ethnic correlations with Mendelian disorders include higher frequencies of cystic fibrosis in whites, sickle cell anemia in blacks, β-thalassemia in Italians and Greeks, α-thalassemia in Asians, and Tay-Sachs disease in Jews.

• Advanced maternal age is associated with higher risks for chromosomal disorders (e.g., Down's syndrome, trisomy 13), while advanced paternal age is associated with higher risks for new mutations (e.g., those producing achondroplasia or Marfan's syndrome).

• The Hardy-Weinberg law predicts allele frequencies in an idealized population according to the formula $p^2 + 2pq + q^2 = 1$. Applied to cystic fibrosis, the law predicts that homozygotes (q^2) have a frequency of 1 in 1600, predicting that carriers ($2pq$) have a frequency of 1 in 20.

• A karyotype is an ordered arrangement of chromosomes that is described by cytogenetic notation. A karyotype can be obtained from dividing cells (blood leukocytes, bone marrow, fibroblasts, amniocytes), but not from frozen or formalin-fixed cells.

• Cytogenetic notation includes the chromosome number (usually 46), description of the sex chromosomes (usually XX or XY), and indication of missing, extra, or rearranged chromosomes. Examples include 47,XY,+21 (male with Down's syndrome); 47,XX,+13 (female with trisomy 13); 45,X (female with monosomy X or Turner's syndrome); 46,XX,del(5p) (female with deletion of the chromosome 5 short arm).

• DNA diagnosis examines specific regions of genes for altered nucleotide sequences or deletions that affect gene expression and function; techniques include Southern blotting, gene amplification with the polymerase chain reaction (PCR), and mutant allele detection by

Pedigree symbols and pedigree patterns.

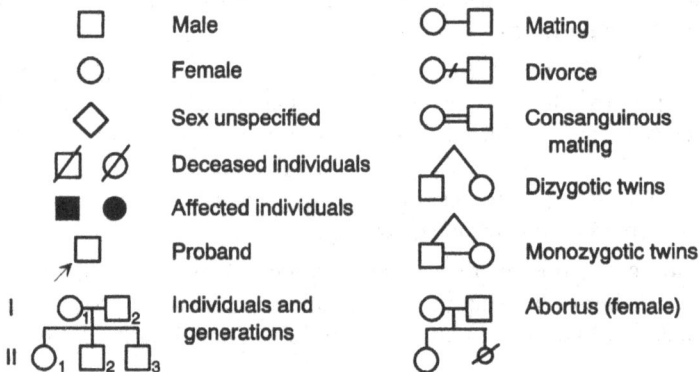

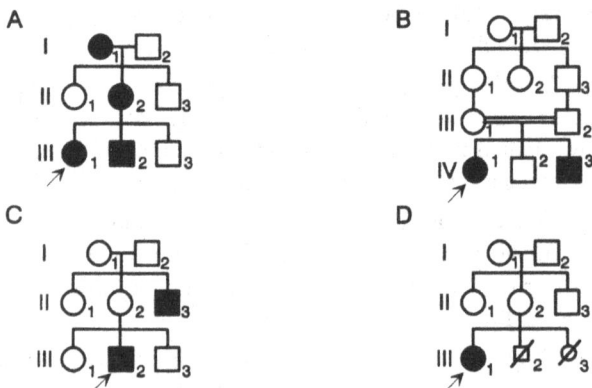

hybridization with allele-specific oligonucleotides (ASOs). Chromosome microdeletions encompass several genes and are detected by fluorescent in situ hybridization (FISH).

• Non-Mendelian inheritance mechanisms include mitochondrial inheritance (exhibiting maternal transmission), expansion of triplet repeats (exhibiting anticipation in pedigrees as in the fragile X syndrome), and genomic imprinting (exhibiting different phenotypes according to maternal or paternal origin of the aberrant genes).

• Prenatal diagnosis can include fetal ultrasound, maternal serum studies, or sampling of cells from the fetoplacental unit by chorionic villus sampling [CVS at 8 to 10 weeks, amniocentesis at 12 to 18 weeks, or percutaneous umbilical sampling (PUBS) from 16 weeks to term].

GENETICALLY BASED BIOCHEMICAL DISEASES

Disease and Incidence*	Defect	Symptoms
Glycolysis-based hemolytic anemias	Deficient glycolytic enzymes	Hemolytic anemia
Glucose-6-P dehydrogenase deficiency (up to 1 in 3)	Deficient enzyme of pentose phosphate shunt	Hemolytic anemia with antimalarial drugs
Glycogen storage diseases (one type XLR)	Deficient glycogen catabolism	Glycogen accumulation
Type 1a von Gierke's disease (1 in 100,000)	Glucose-6-phosphatase deficiency	Large liver, hypoglycemia
Type II Pompe's disease (1 in 100,000)	Lysomal α-glucosidase deficiency	Short PR interval on ECG, fatal
Type III Cori's disease	Debranching enzyme deficiency	Large liver, mild myopathy
Type V McArdle's disease (1 in 100,000)		
Lipid storage diseases	Deficiencies of sphingolipid metabolism	Sphingolipid storage neurodegeneration
Tay-Sachs disease [1 in 4,000 (Jews); 1 in 100,000]	Hexosaminidase A deficiency	Cherry red spot, neurodegeneration
Krabbe's disease (1 in 100,000)	Galactosylceramide β-galactosidase deficiency	Neurodegeneration, demyelination
Niemann-Pick disease type A (1 in 50,000)	Sphingomyelinase deficiency	Organomegaly, neurodegeneration
Gaucher's disease type I [1 in 1,000 (Jews); 1 in 100,000]	Glucosylceramide β-glucosidase deficiency	Organomegaly, fractures
Fabry's disease (XLR) (1 in 40,000)	β-galactosidase deficiency	Angiokeratoma, nerve pains
Lipid transport diseases (most AD)	Abnormality in plasma lipoprotein receptors or enzymes	Fatty serum, atherosclerosis

GENETICALLY BASED BIOCHEMICAL DISEASES (*CONT.*)

Disease and Incidence*	Defect	Symptoms
Familial hypercholesterolemia type IIa [1 in 500 (AD)]	Defective apo-B100 LDL receptors	Xanthomas, hypercholesterolemia
Familial hypertriglyceridemia type IV [1 in 50,000 (AD)]	Increased synthesis or decreased catabolism of VLDLs	Hypertriglyceridemia, atherosclerosis
Ion channel diseases (cystic fibrosis)	Sodium transport deficiency	High sweat chloride, lung disease
Nucleotide catabolism diseases (some XLR)	PRPP synthetase abnormalities	Gouty arthritis
Lesch-Nyhan syndrome [(XLR) 1 in 100,000]	Deficient HGPRT	Self-mutilation
DNA repair diseases (xeroderma pigmentosum)	Exonuclease deficiency	Skin cancer
Hereditary nonpolyposis colorectal cancer [(AD) 1 in 50,000]	DNA mismatch repair defects	Colon cancer
RNA-processing diseases [thalassemias (1 in 50,000; higher in Mediterraneans (β) or Asians (α)]	Imbalance of α- or β-hemoglobin chains	Anemia
Porphyrias [(one form AD) 1 in 1 million]	Heme biosynthesis enzyme defects	Abdominal pain, psychosis, skin rash
Amino acid metabolism diseases [Phenylketonuria (1 in 12,000)	Phenylalanine hydroxylase deficiency	Mousy odor, pale skin, blond hair
Maple syrup urine disease (1 in 100,000)]	Branched-chain amino acid dehydrogenase deficiency	Seizures, acidosis

*All diseases are autosomal recessive unless otherwise indicated. AD, autosomal dominant; XLR, X-linked recessive; PRPP, 5-phosphoribosyl-1-pyrophosphate; HGPRT, hypoxanthine-guanine phosphoribosyl-transferase.

High-Yield Facts
in Microbiology

- Detection of HIV RNA by nucleic acid amplification of the viral load is the best predictor of "progression to AIDS." (Virology)

- HIV RNA PCR and sequencing of the amplified products may be used to monitor resistance to anti-HIV drugs. HIV patients with total CD4 lymphocyte counts of less than 200 CD4 cells/μL are susceptible to opportunistic infections such as those caused by *Cryptococcus, Mycobacterium,* and *Pneumocystis.* (Virology)

- *Cyclospora* is an ooidian parasite similar to *Cryptosporidium.* It causes food-borne diarrheal illness and has been associated with contaminated berries. (Parasitology)

- *Giardia,* a large flagellate with both cyst and trophozoite forms, is the most common parasitic disease in the United States. The disease is characterized by diarrhea, cramping, and fever. (Parasitology)

- Enterohemorrhagic *E. coli* causes bloody diarrhea and hemolytic uremic syndrome. The mode of action is production of Shiga-like toxin by *E. coli.* (Bacteriology)

- Vancomycin-resistant enterococci, methicillin-resistant *Staphylococcus aureus* (MRSA), and vancomycin-indeterminate *S. aureus* (VISA) are among the most feared nosocomial pathogens. A recently introduced antibiotic, quinapristin-delfapristin, effectively treats vancomycin-resistant enterococci or the few vancomycin-indeterminate MRSAs that have occurred. (Bacteriology)

- Following an upsurge of tuberculosis in the early 1990s, cases of *Mycobacterium tuberculosis* infection have remained static. *M. tuberculosis* causes initial primary pulmonary infection as well as a chronic disease characterized by hemoptysis, loss of weight, and fever. (Bacteriology)

- Penicillin-resistant pneumococci (*Streptococcus pneumoniae*) may account for up to 40% of isolates of *S. pneumoniae.* Third- or fourth-generation cephalosporins may be used as alternative treatment as well as vancomycin and rifampin. (Bacteriology)

- *Ehrlichia,* a recently emerging tick-borne pathogen, is transmitted by *Ixodes scapularis,* the same tick that transmits the Lyme disease bacterium. *Ehrlichia* is also transmitted by the Lone Star tick, *Amblyomma americanum.* (*Chlamydia, Rickettsia*)

- Eastern equine encephalitis may be transmitted to humans by the bite of a mosquito, particularly in the northeastern United States. (Virology)

- Transfusion-associated babesiosis is a growing problem, particularly in the immunosuppressed or patients without a spleen. Tick-borne babesiosis caused by the same tick that transmits Lyme disease is an emerging infection. (Parasitology)

- Dengue fever, a viral illness transmitted by the *Aedes* mosquito, is prevalent in epidemic proportions in both the Caribbean and Southeast Asia. (Virology)

- There are five major classes of immunoglobulin: IgG, IgM, IgA, IgD, IgE. These immunoglobulins are distinguished by differences in the C regions of each individual H chain. These differences are function-related. (Immunology)

- Peptidoglycans are unique to prokaryotic organisms. They consist of a glycan backbone of muramic acid and glucosamine as well as cross-linked peptides. The enzymes responsible for cross-linking (transpeptidases) are the targets for β-lactam antibiotics. (Physiology)

- Genetic exchange in microorganisms occurs by several mechanisms, including transformation, transduction, conjugation, and transposition. These processes are the basis for gene cloning in microorganisms. (Physiology)

- Virulence factors in bacteria include adherence factors, invasins, capsules, endotoxin, and exotoxin. Such factors enable microorganisms to invade the host, cause disease, and resist host defense mechanisms. (Physiology)

- Sites of action of antimicrobial agents include cell-wall synthesis, cell membrane integrity, DNA replication, protein synthesis, DNA-dependent RNA polymerase, and folic acid metabolism. (Physiology)

- *Staphylococcus aureus* expresses two types of superantigens: enterotoxin (responsible for staphylococcal food poisoning) and toxic shock toxin. (Bacteriology)

- Free radicals of oxygen (superoxides) kill anaerobic bacteria exposed to air. Superoxide dismutase is a potent bacterial antioxidant. The presence of peroxidases in bacteria are protective. (Physiology)

- *Campylobacter* and *Helicobacter* are both helical-shaped bacteria. *Helicobacter* is known to play a role in the pathogenesis of peptic ulcer disease, while *Campylobacter* causes a food-borne gastrointestinal illness, most commonly from undercooked meat. Both bacteria are susceptible to antibiotics such as tetracycline. *Helicobacter* may be treated with Pepto-Bismol, metronidazole, and amoxicillin. (Bacteriology)

- The agents of bovine spongiform encephalopathy (Mad Cow Disease), scrapies, and new-variant Creutzfeldt-Jakob disease in humans are *prions* or amyloid fibrils. Also included are prions that cause chronic wasting disease (CWD) in elk and deer, although these agents of CWD have not been shown to be transmissable to either cattle or humans. These self-replicating proteins are resistant to heat and chemical agents. (Virology)

- Prior to 1999, West Nile virus, an arbovirus with serological cross-reactivity to St. Louis encephalitis virus was not seen in the United States. However, during 1999 and 2000, a large number of birds were infected with West Nile virus, as well as a few humans, some of whom died. (Virology)

- The genotype of hepatitis C is important in predicting the response of this virus to therapy with interferon and ribavirin as well as the required length of treatment. (Virology)

High-Yield Facts in Neuroscience

GROSS ANATOMY OF THE BRAIN

1. Lateral view of the brain. The loci of key motor and sensory structures of the cerebral cortex are indicated in this figure. Anatomical definitions: anterior—toward the front (rostral end) of the forebrain; posterior—toward the back (caudal end) of the forebrain; dorsal—toward the superior surface of the forebrain; ventral—toward the inferior surface of the forebrain. Note that with respect to the brainstem and spinal cord, the terms *anterior* and *ventral* are synonymous; likewise, *posterior* and *dorsal* are also synonymous. Here, the term *rostral* means toward the midbrain, and the term *caudal* means toward the sacral aspect of spinal cord.

2. Midsagittal view of the brain. Magnetic resonance image: T2-weighted, high-resolution, fast spin echo image.

3. Horizontal (transaxial) view of the brain. Magnetic resonance image: Fast inversion recovery for myelin suppression image.

4. Frontal view of the brain. Magnetic resonance image: Fast inversion recovery for myelin suppression image.

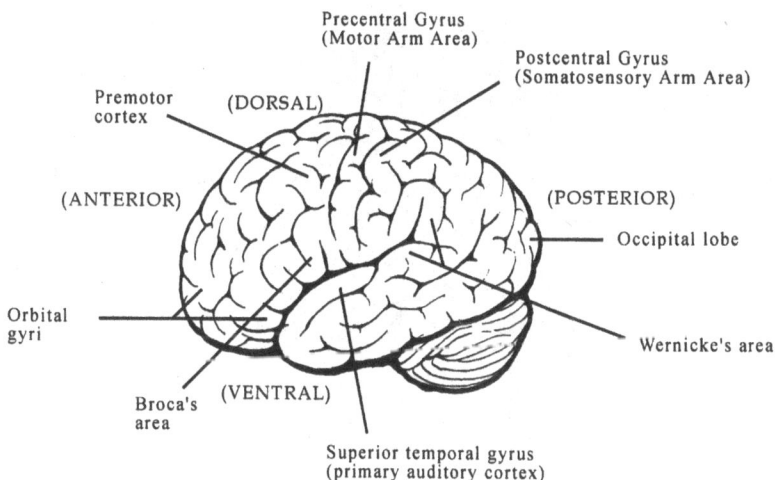

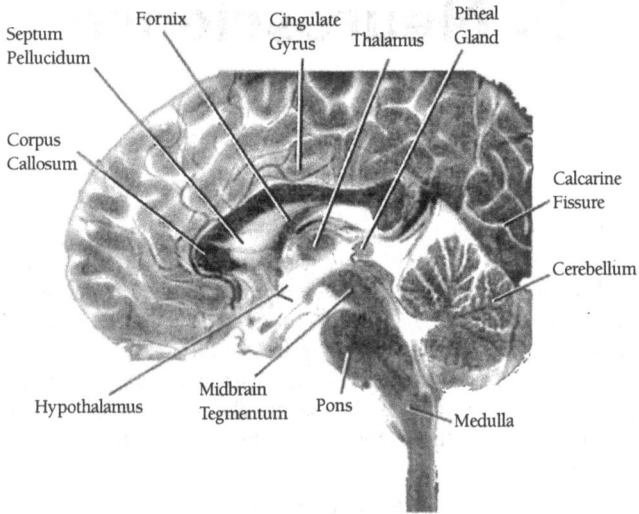

Septum Pellucidum
Fornix
Cingulate Gyrus
Thalamus
Pineal Gland
Corpus Callosum
Calcarine Fissure
Cerebellum
Hypothalamus
Midbrain Tegmentum
Pons
Medulla

(Courtesy of Leo J. Wolansky, M.D.)

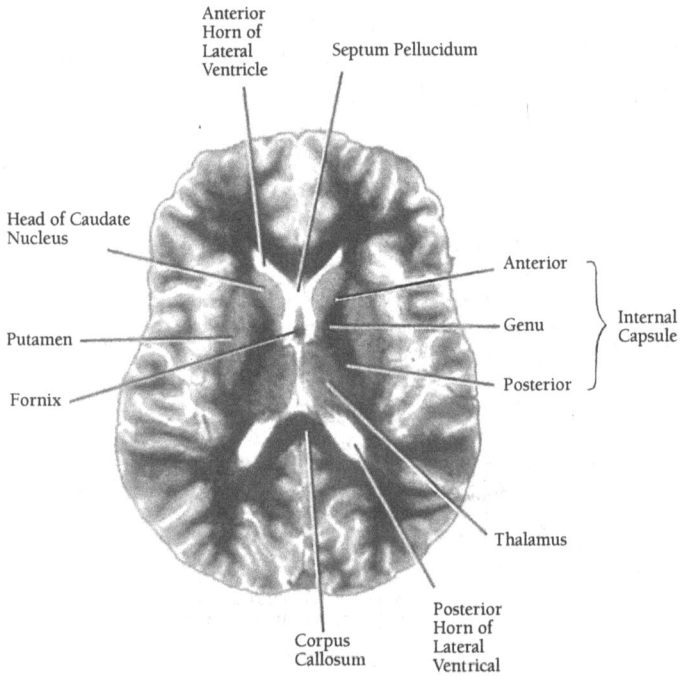

Anterior Horn of Lateral Ventricle
Septum Pellucidum
Head of Caudate Nucleus
Anterior
Internal Capsule
Putamen
Genu
Fornix
Posterior
Thalamus
Corpus Callosum
Posterior Horn of Lateral Ventrical

(Courtesy of Leo J. Wolansky, M.D.)

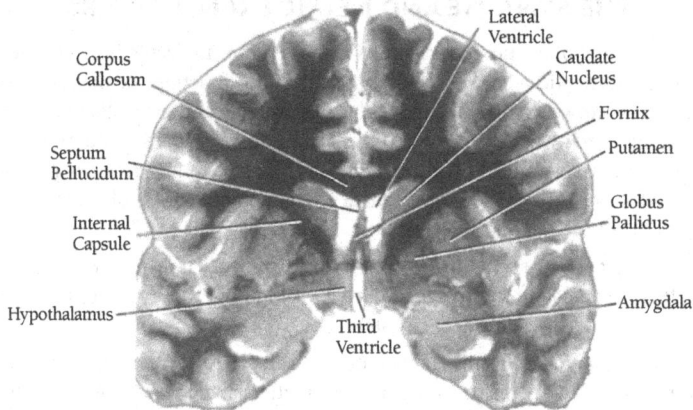

Corpus Callosum

Septum Pellucidum

Internal Capsule

Hypothalamus

Lateral Ventricle

Caudate Nucleus

Fornix

Putamen

Globus Pallidus

Amygdala

Third Ventricle

(Courtesy of Leo J. Wolansky, M.D.)

DEVELOPMENT

5. The sulcus limitans divides the alar plate, from which sensory regions of the spinal cord and brainstem are formed, and a basal plate, from which motor regions of the spinal cord and brainstem are formed.

THE NEURON

6. The neuron consists of a cell body, dendrites (which extend from the cell body), and an axon. Activation of sodium channels is associated with membrane depolarization, while activation of potassium and chloride channels is associated with membrane hyperpolarization. After information is received from a presynaptic neuron, depolarization occurs in the postsynaptic neuron; then, the action potential is initiated and propagated down the axon from the initial segment.

7. Myelin formation is produced in the peripheral nervous system by numerous Schwann cells, while a similar function in the central nervous system is carried out by an oligodendrocyte, which can wrap itself around numbers of neurons. Myelination in the nervous system allows for rapid conduction of action potentials by a process of saltatory conduction, in which the signals skip along openings in the myelin called *nodes of Ranvier*. Neurons that are myelinated (e.g., the pyramidal tracts and dorsal column-medial lemniscal system) are rapidly conducting, whereas those that are poorly or nonmyelinated (e.g., certain pain-afferent fibers to the spinal cord) are slowly conducting. Damage to such myelinated neurons typically disrupts the transmission of neural signals and is frequently seen in autoimmune diseases such as multiple sclerosis, in which sensory and motor functions are severely compromised.

THE SYNAPSE AND NEUROTRANSMITTERS

8. The binding of the neurotransmitter to the receptor molecule is determined by the postsynaptic receptor, which serves a gating function for particular ions. The receptor is responsible for opening or closing ligand-gated channels, which are regulated by noncovalent binding of compounds such as neurotransmitters. The neurotransmitter, which is contained in presynaptic vesicles and released onto the postsynaptic terminal, causes activation of the receptor, which in turn, produces postsynaptic potentials.

9. The sequence of events in synaptic transmission is: transmitter synthesis → release of transmitter into synaptic cleft → binding of transmitter to postsynaptic receptor → removal of transmitter.

10. Major excitatory transmitters include: Substance P, acetylcholine, and excitatory amino acids; major inhibitory transmitters include: GABA, enkephalin, and glycine. Disruption of neurotransmitter function can lead to different diseases of the nervous system. One such example involves the role of acetylcholine at the neuromuscular junction. When antibodies are formed against the acetylcholine receptor at the neuromuscular junction, transmission is disrupted and the autoimmune disease called *myasthenia gravis* occurs. This disorder includes symptoms such as weakness and fatigue of the muscles.

SPINAL CORD

11. Major ascending tracts of the spinal cord and their functions include:

Dorsal columns. Mediates conscious proprioception, two-point discrimination, and some tactile sensation ipsilaterally to the dorsal column nuclei and then contralaterally from the dorsal column nuclei to the postcentral gyrus from the VPL of the thalamus.

Lateral spinothalamic tract. Mediates pain and temperature inputs contralaterally to the VPL and posterior complex of the thalamic nuclei and then to the postcentral gyrus.

Anterior spinothalamic tract. Mediates tactile impulses contralaterally to the VPL and then to the postcentral gyrus.

Posterior spinocerebellar tract. Mediates unconscious proprioception from muscle spindles and Golgi tendon organs of the lower limbs through the inferior cerebellar peduncle ipsilaterally to the anterior lobe of the cerebellar cortex.

Cuneocerebellar tract. Mediates unconscious proprioception from muscle spindles and Golgi tendon organs of the upper limbs from the accessory cuneate nucleus through the inferior cerebellar peduncle to the anterior lobe of the cerebellar cortex.

Anterior spinocerebellar tract. Mediates unconscious proprioception from the Golgi tendon organs of the lower limbs bilaterally to the anterior lobe of the cerebellar cortex. This tract initially crosses in the spinal cord and then crosses again through the superior cerebellar peduncle.

Major descending tracts of the spinal cord and their functions include:

Lateral corticospinal tract. Mediates voluntary control of motor functions from the contralateral cerebral cortex to all levels of the spinal cord.

Rubrospinal tracts. Mediates descending excitation of flexor motor neurons at both the cervical and lumbar levels of the contralateral spinal cord.

Reticulospinal tracts. The lateral reticulospinal tract arises from the medulla and descends bilaterally to the cervical and lumbar levels of the spinal cord, mediating inhibition upon the spinal reflexes, mainly of extensors; the medial reticulospinal tract arises from the pons and descends mainly ipsilaterally to the cervical and lumbar levels of the spinal cord and facilitates extensor reflexes.

Vestibulospinal tracts. The lateral vestibulospinal tract arises from the lateral vestibular nucleus and descends ipsilaterally to the cervical and lumbar levels of the spinal cord, mediating powerful excitation of the extensor motor neurons; the medial vestibulospinal tract arises from the medial vestibular nucleus and descends mainly to the cervical levels of the spinal cord, mediating postural reflexes of the head and neck.

12. Major disorders of the spinal cord include:

Brown-Séquard's syndrome. Hemisection of the spinal cord often due to a bullet or knife wound—contralateral loss of pain and temperature below the level of the lesion; bilateral segmental loss of pain and temperature at the level of the lesion; ipsilateral loss of conscious proprioception below the level of the lesion; ipsilateral upper motor neuron paralysis below the level of the lesion; ipsilateral lower motor neuron paralysis at the level of the lesion.

Tabes dorsalis. Damage to the dorsal root ganglion and dorsal columns resulting from syphilis—ipsilateral loss of conscious proprioception and tendon reflexes.

Amyotrophic lateral sclerosis (ALS). A disease whose etiology is not yet known that destroys both corticospinal fibers and ventral horn cells, causing abnormal reflexes, muscle weakness, atrophy, and ultimately death.

Syringomyelia. Caused by abnormal closure of the central canal during development, by trauma, or by a tumor, the result of which is an enlargement of the central canal, causing a segmental bilateral loss of pain and temperature due to damage to the decussating spinothalamic fibers.

Combined systems disease. Results from pernicious anemia associated with a deficiency in vitamin B_{12}; there is degeneration of both the dorsal columns and corticospinal tracts, resulting in a loss of conscious proprioception, position sense, upper motor neuron symptoms, and muscle weakness.

AUTONOMIC NERVOUS SYSTEM

13. The sympathetic nervous system arises from the thoracic and lumbar cords (T_1-L_2), and the parasympathetic nervous system arises from S_2-S_4 and cranial nerves III, VII, IX, and X. All preganglionic neurons are cholinergic as well as parasympathetic postganglionic neurons. In addition, sympathetic postganglionic innervation of sweat glands and blood vessels in skeletal muscle is also cholinergic. Most other postganglionic sympathetic endings are adrenergic. Examples of functions of the sympathetic nervous system include: pupillary dilation, acceleration of heart rate, constriction of blood vessels of the trunk and extremities, and inhibition of gastric motility. Examples of functions of the parasympathetic nervous system include: pupillary constriction, decrease in heart rate, secretion of the salivary and lacrimal glands, and stimulation of gastric motility.

THE BRAINSTEM AND CRANIAL NERVES

14. *Lateral medullary syndrome (Wallenberg's syndrome).* Lesions of the lateral aspect of the lower half of the brainstem, due to occlusion of the inferior cerebellar arteries, produce loss of pain and temperature on the same side of the face and opposite side of the body, as well as Horner's syndrome (i.e., myosis, ptosis, and decreased sweating on one side of the face due to disruption of the sympathetic supply to the orbit and pupil, or, as applies in the present context, to disruption of descending sympathetic fibers through the brainstem to the spinal cord).

15. *Medial medullary syndrome.* Lesions of the medial aspect of the medulla typically resulting from occlusion of the anterior spinal artery produce contralateral loss of conscious proprioception, contralateral hemiparesis, and weakness of tongue muscles, which are protruded to the side of the lesion. Body paralysis, which involves the side contralateral to the lesion, coupled with cranial nerve weakness, which involves the side ipsilateral to the lesion, is called *alternating hypoglossal hemiplegia.*

16. Cranial nerves mediate multiple functions in the nervous system: motor nuclei—general somatic efferent (NIII, NIV, NVI, and NXII), special visceral efferent (NV, NVII, NIX, NX, and NXI), general visceral efferent (NIII, NVII, NIX, and NX), general somatic afferent (NV, NIX, and NX), special sensory afferent (NII and NVIII), or special visceral afferent (NI, NVII, NIX, and NX).

SENSORY SYSTEMS

17. Loss of partial or total aspects of the visual field can be understood in terms of damage to the retinal pathways, including their targets in the lateral geniculate nucleus and visual cortex. The schematic diagram depicts the kinds of field deficits that occur following lesions of different aspects of the visual pathway. Key: (A) optic nerve lesion producing total blindness in the left eye; (B) lesion that disrupts the right retinal nasal fibers that project from the base of the left optic nerve producing right upper quadrantanopia and left scotoma; (C) lesion of optic chiasm producing bitemporal hemianopsia; (D) unilateral (left) optic tract lesion producing a right homonymous hemianopsia; (E) interruption of left visual radiations that pass ventrally through the temporal lobe to the lower bank of the visual cortex (i.e., the loop of Meyer) producing an upper right

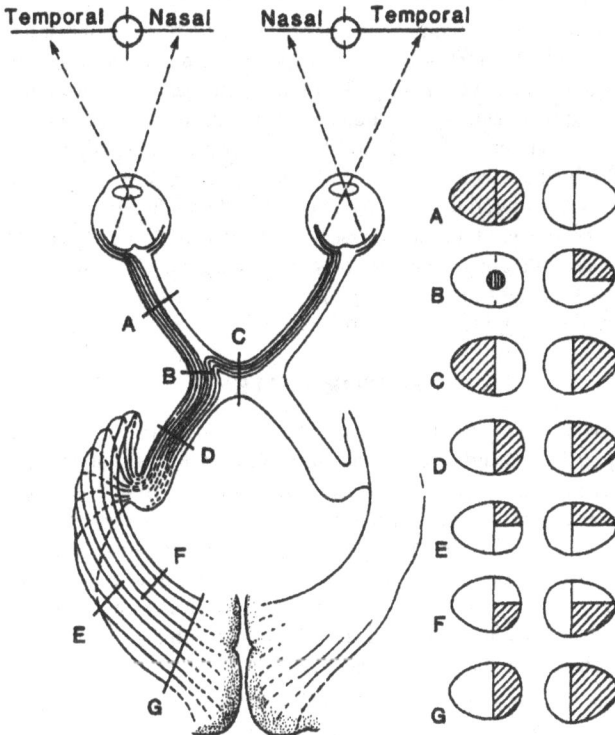

(Reproduced, with permission, from Adams RD, Victor M, Ropper AH: Principles of Neurology, 6/e. New York, McGraw-Hill, 1997.)

quadrantanopia; (F) interruption of left visual radiations that pass more dorsally through the occipital lobe to the upper bank of the visual cortex, producing a lower right quadrantanopia; and (G) lesion of the left visual cortex that produces a right homonymous hemianopsia.

18. The principles of an excitatory focus-and-surround inhibition, as well as that of a somatotropic organization are present within a given receptor system and form the functional basis for discriminative functions in a number of the sensory systems, including the auditory circuit. The auditory pathways are complex and involve the following synaptic connections: first-order root fibers of the spiral ganglion, which originate in the cochlea (organ of Corti), synapse in the cochlear nuclei of the upper medulla; second-order neurons, which project through lateral lemnisci, terminate bilaterally in the inferior colliculus; third-order neurons project to the medial geniculate nucleus; and fourth-order neurons project to the superior temporal gyrus (primary auditory cortex).

19. *Vestibular pathway.*   First-order neurons originate from vestibular ganglia and have peripheral processes located in specialized receptors in the utricle, saccule, and semicircular canals. The central branches of this neuron reach the brain and terminate in the vestibular nuclei. Second-order neurons may pass to the cerebellar cortex (flocculonodular lobe) or project directly into the medial longitudinal fasciculus, where the fibers may run in a rostral or caudal direction, terminating in the NIII, NIV, or NVI of the midbrain and pons or spinal cord, respectively. Damage to these fibers, especially within the medial longitudinal fasciculus or cerebellum, produces nystagmus (i.e., involuntary movement of the eyes, in the horizontal or vertical plane, first slowly and then followed by a rapid jerking return).

MOTOR SYSTEMS

20. Voluntary motor control affecting mainly the flexor system is expressed through the descending pyramidal tracts plus the rubrospinal tract, and control of functions associated with posture is mediated through such descending pathways as the vestibulo- and reticulospinal tracts. Modulation of motor functions is mediated by the basal ganglia and cerebellum. Involuntary motor disturbances at rest (called *dyskinesias*) are associated with the disruption of functions of the basal ganglia, and motor disturbances occurring during attempts at movement are frequently associated with damage to the cerebellum or its afferent or efferent pathways.

A. The basal ganglia consist of the neostriatum (caudate nucleus and putamen), the paleostriatum (globus pallidus), and two additional structures that are anatomically and functionally related to the basal ganglia—the substantia nigra and subthalamic nucleus.

1. Disorders of the basal ganglia

   a. *Parkinson's disease.* Characterized by "pill rolling" tremor, akinesia (poverty of movement), and rigidity. This disorder is due to a reduction in striatal dopamine following a loss of dopamine neurons in the pars compacta of the substantia nigra, which project to the neostriatum.

   b. *Chorea.* Characterized by short, jerky movements of the distal extremities at rest. It is associated with lesions of the striatum. One form of chorea, called *Huntington's chorea,* is a genetic disorder that is associated with a chromosomal mutation. It results in destruction of GABAergic and cholinergic neurons in the caudate nucleus, and there is concomitant loss of neurons in the prefrontal regions of the neocortex.

   c. *Athetosis.* Characterized by slow writhing movements of the extremities and muscles of the neck. The lesion may involve the striatum.

   d. *Hemiballism.* Characterized by wild (flailing) movements of the limbs on one side of the body. It is due to damage of the subthalamic nucleus on the contralateral side.

2. Disorders of the cerebellum

   a. *Anterior lobe (paleocerebellum).* Characterized by a wide, staggering gait ataxia resulting primarily from damage that affects the vermal and paravermal regions of the anterior lobe.

   b. *Posterior lobe (neocerebellum).* Characterized most frequently by loss of coordination while executing voluntary movements.

   c. *Flocculonodular lobe (archicerebellum).* Characterized by a loss of equilibrium with the patient displaying a wide, staggering ataxic gait. Lesions of this region also produce eye movement disorders, including nystagmus.

HIGHER AUTONOMIC AND BEHAVIORAL FUNCTIONS

21. Control of autonomic and endocrine functions as well as emotional behavior are mediated by the limbic system, hypothalamus, and midbrain periaqueductal gray matter. Interrelationships among these three groups of structures are as follows:

   A. For the expression of emotional behavior, such as rage and autonomic functions, these are mediated from the medial hypothalamus → midbrain periaqueductal gray → autonomic (i.e., neurons in lower medulla that regulate heart rate, blood pressure, and respiration) and somatomotor neurons (i.e., neurons of the trigeminal nerve that control vocalization), autonomic, and somatomotor neurons of the spinal cord. These processes are further regulated by different groups of neurons within the limbic system (i.e., hippocampal formation, amygdala, septal area), which produce their effects by projecting directly or indirectly to the hypothalamus or midbrain periaqueductal gray.

B. For the regulation of endocrine functions, these are mediated from the supraoptic and paraventricular nuclei of the hypothalamus → posterior lobe of the pituitary, and from the medial hypothalamus → anterior lobe of the pituitary (via the vascular system). Limbic projections to the hypothalamus enable structures such as the hippocampal formation, amygdala, and septal area to modulate endocrine functions of the hypothalamus. Disruption of hypothalamic neurons may alter the mechanism for the expression of rage behavior and, likewise, affect temperature regulation, sexual behavior, feeding, drinking, and endocrine functions. Damage to neurons of the limbic system frequently leads to temporal lobe epilepsy and changes in the threshold for the expression of rage behavior [i.e., when different groups of neurons in the amygdala are damaged, heightened aggressiveness or a reduction in aggression (such as the Klüver-Bucy syndrome) may ensue]. Damage to the hippocampal formation can result in temporal lobe epilepsy and short-term memory deficits.

CEREBRAL CORTEX

22. Dysfunctions associated with cerebrovascular accidents and tumors can be understood in terms of the principles of cortical localization and cerebral dominance. Following is a list of common disorders, their descriptions, and the cortical regions most closely associated with each disorder.

A. *Upper motor neuron paralysis.* Damage to the precentral, premotor, and supplementary motor areas (as well as the internal capsule, crus cerebri, or corticospinal tracts), resulting in a loss of voluntary control of the upper and lower limbs, depending upon the extent of the lesion. This disorder is also associated with hyperreflexia, hypertonicity, and a positive Babinski's sign.

B. *Broca's aphasia.* Damage to the inferior frontal gyrus of the dominant hemisphere. The patient cannot name simple objects but has no difficulty in comprehending spoken language.

C. *Wernicke's aphasia.* Damage to the region of the superior temporal gyrus and/or adjoining regions. The patient has difficulty in comprehending language but speech appears fluent.

D. *Astereognosia.* Damage to the parietal cortex of the contralateral side results in a failure of tactile recognition of objects (e.g., a blackboard eraser, a pack of cigarettes).

E. *Unilateral sensory neglect.*   Damage to the parietal lobe (usually of the right hemisphere) can cause this disorder. The patient typically ignores stimuli on the opposite (i.e., left) side of body space, which includes visual, somatosensory, and auditory stimuli. The individual will neglect the opposite side of the body by neglecting to shave that side of the face and by denying that there is anything wrong with that side of the body, which may include a motor paralysis. The patient may further draw a picture of flowers or of a clock in which the petals on the flowers or the numbers on the clock are limited to the right side of each of the figures.

F. *Apraxia.*   Damage to the posterior parietal cortex can prevent an individual from conceptualizing the sequence of events necessary to carry out a task, even though the basic sensory and motor pathways necessary to produce the required movements are intact. In effect, the patient is thus unable to carry out the task.

High-Yield Facts in Pathology

I. CELL INJURY

Reversible Cell Injury
- swelling of cell organelles and entire cell
- dissociation of ribosomes from endoplasmic reticulum
- decreased energy production by mitochondria
- increased glycolysis → decreased pH → nuclear chromatin clumping

Irreversible Cell Injury
- dense bodies within mitochondria (flocculent densities in heart)
- release of cellular enzymes (e.g., SGOT, LDH, and CPK after MI)
- nuclear degeneration (pyknosis, karyolysis, karyorrhexis)
- cell death

2. FATTY CHANGE OF THE LIVER

Mechanisms
1. Increased delivery of free fatty acids to liver
   - starvation
   - corticosteroids
   - diabetes mellitus
2. Increased formation of triglycerides
   - alcohol (note: NADH > NAD)
3. Decreased formation of apoproteins
   - carbon tetrachloride
   - protein malnutrition (kwashiorkor)

3. CELL DEATH

Apoptosis
- "programmed" cell death
- single cells (not large groups of cells)
- cells shrink → form apoptotic bodies
- gene activation → forms endonucleases
- peripheral condensation of chromatin with DNA ladder
- no inflammatory response

Examples of apoptosis:
1. Physiologic
   - involution of thymus
   - cell death within germinal centers of lymph nodes
   - fragmentation of endometrium during menses
   - lactating breast during weaning
2. Pathologic
   - viral hepatitis
   - cytotoxic T cell–mediated immune destruction (type IV hypersensitivity)

Necrosis
- cause → hypoxia or toxins (irreversible injury)
- many cells or clusters of cells
- cells swell
- inflammation present

Examples of necrosis:
- coagulative necrosis → ischemia (except the brain)
- liquefactive necrosis → bacterial infection (and brain infarction)
- fat necrosis → pancreatitis and trauma to the breast
- caseous necrosis → tuberculosis
- fibrinoid necrosis → autoimmune disease (type III hypersensitivity reaction)
- gangrene → ischemia to extremities → dry (mainly coagulative necrosis) or wet (mainly liquefactive necrosis due to bacterial infection)

4. TERMS
Adaptation
- hypertrophy → increase in the size of cells
- hyperplasia → increase in the number of cells
- atrophy → decrease in the size of an organ
- aplasia → failure of cell production
- hypoplasia → decrease in the number of cells
- metaplasia → replacement of one cell type by another
- dysplasia → abnormal cell growth

Abnormal Organ Development
- anlage → primitive mass of cells
- aplasia → complete failure of an organ to develop (anlage present)

- agenesis → complete failure of an organ to develop (no anlage present)
- hypoplasia → reduction in the size of an organ due to a decrease in the number of cells
- atrophy → decrease in the size of an organ due to a decrease in the number of preexisting cells

5. CARDINAL SIGNS OF INFLAMMATION

- rubor → red
- calor → hot
- tumor → swollen
- dolor → pain

6. COMPLEMENT CASCADE

Products
- C3b → opsonin
- C5a → chemotaxis and leukocyte activation
- C3a, C4a, C5a → anaphylatoxins
- C5–9 → membrane attack complex

Deficiencies
- deficiency of C3 and C5 → recurrent pyogenic bacterial infections
- deficiency of C6, C7, and C8 → recurrent infections with *Neisseria* species
- deficiency of C1 esterase inhibitor → hereditary angioedema
- deficiency of decay-accelerating factor → paroxysmal nocturnal hemoglobinuria

7. THROMBOXANE VS. PROSTACYCLIN

Thromboxane
- produced by platelets
- causes vasoconstriction
- stimulates platelet aggregation

Prostacyclin
- produced by endothelial cells
- causes vasodilation
- inhibits platelet aggregation

8. GRANULOMATOUS INFLAMMATION

Caseating Granulomas
- aggregates of activated macrophages (epitheloid cells)
- tuberculosis

Noncaseating Granulomas
- sarcoidosis
- fungal infections
- foreign-body reaction

9. COLLAGEN TYPES

Fibrillar Collagens
- type I → skin, bones, tendons, mature scars
- type II → cartilage
- type III → embryonic tissue, blood vessels, pliable organs, immature scars

Amorphous Collagens
- type IV → basement membranes
- type VI → connective tissue

10. EDEMA

Exudates
1. Composition
   - increased protein
   - increased cells
   - specific gravity greater than 1.020
2. Cause
   - inflammation
   - increased blood vessel permeability

Transudates
1. Composition
   - no increased protein
   - no increased cells
   - specific gravity less than 1.012
2. Cause → abnormality of Starling forces
   a. increased hydrostatic (venous) pressure
      - congestive heart failure
      - portal hypertension
   b. decreased oncotic pressure → due to decreased albumin
      - liver disease
      - renal disease (nephrotic syndrome)

c. lymphatic obstruction
  • tumors or surgery
  • filaria

11. CARCINOMAS

Squamous Cell Carcinoma
• skin cancer
• lung cancer
• esophageal cancer
• cervical cancer

Adenocarcinoma
• lung cancer
• colon cancer
• stomach cancer
• prostate cancer
• endometrial cancer

Transitional Cell Carcinoma
• urinary bladder cancer
• renal cancer (renal pelvis)

Clear Cell Carcinoma
• renal cortex
• vaginal cancer (associated with DES)

12. NEOPLASMS

Benign
• grow slowly
• remain localized
• may have well-developed fibrous capsule
• do not metastasize
• well differentiated histologically

Malignant
• grow rapidly
• locally invasive
• irregular growth; no capsule
• capable of metastasis
• variable degrees of differentiation (well differentiated, moderately differentiated, poorly differentiated)

13. ONCOGENE EXPRESSION

Growth Factors

1. *c-sis*
   - β chain of platelet-derived growth factor
   - astrocytomas and osteogenic sarcomas

Growth Factor Receptors

1. *c-erb B1*
   - receptor for epidermal growth factor
   - breast cancer and squamous cell carcinoma of the lung
2. *c-neu*
   - receptor for epidermal growth factor
   - breast cancer
3. *c-fms*
   - receptor for colony-stimulating factor (CSF)
   - leukemia

Abnormal Membrane Protein Kinase

1. *c-abl*
   - membrane tyrosine kinase
   - chronic myelocytic leukemia (CML)

GTP-Binding Proteins

1. *c-ras*
   - product is p21 (protein)
   - adenocarcinomas

Nuclear Regulatory Proteins

1. *c-myc* → Burkitt's lymphoma
2. *N-myc* → neuroblastoma
3. *L-myc* → small cell carcinoma of the lung
4. *c-jun*
5. *c-fos*

14. CHROMOSOMES AND CANCER

Point Mutations

- *c-ras* → adenocarcinomas

Translocations
- *c-abl* on chromosome 9 → CML
- *c-myc* on chromosome 8 → Burkitt's lymphoma
- *bcl-2* on chromosome 18 → nodular lymphoma

Gene Amplification
- *N-myc* → neuroblastoma
- *c-neu* → breast cancer
- *c-erb B2* → breast cancer

15. ANTIONCOGENES

Tumor Suppressor Genes
- Rb → retinoblastoma and osteogenic sarcoma
- p53 → many tumors and the Li-Fraumeni syndrome
- WT1 → Wilms' tumor and aniridia
- NF1 → neurofibromatosis type 1

16. CHEMICAL CARCINOGENS

Initiators
- tobacco smoke → many tumors
- benzene → leukemias
- vinyl chloride → angiosarcomas of the liver
- β-naphthylamine → cancer of the urinary bladder
- azo dyes → tumors of the liver
- aflatoxin → hepatoma
- asbestos → mesotheliomas and lung tumors
- arsenic → skin cancer

Promoters
- saccharin → bladder cancer in rats
- hormones (estrogen)

17. VIRUSES AND CANCER

RNA Viruses
- acute-transforming viruses
- slow-transforming viruses
- HTLV-1 → adult T cell leukemia/lymphoma

DNA Viruses

1. HPV (different subtypes)
   - cervical neoplasia
   - condyloma
   - verruca vulgaris
2. EBV
   - African Burkitt's lymphoma
   - carcinoma of the nasopharynx
   - B cell immunoblastic lymphoma
3. Hepatitis B and hepatitis C
   - liver cancer

18. PARANEOPLASTIC SYNDROMES

- Cushing's syndrome (increased cortisol) → lung cancer
- carcinoid syndrome (increased serotonin) → lung cancer or carcinoid tumor of the small intestine
- syndrome of inappropriate ADH secretion (SIADH) → lung cancer and intracranial neoplasms
- hypercalcemia → lung cancer or multiple myeloma
- hypocalcemia → medullary carcinoma of the thyroid (secretes procalcitonin; stains as amyloid)
- hypoglycemia → liver cancer and tumors of the mesothelium (mesotheliomas)
- polycythemia (erythropoietin) → kidney tumors, liver tumors, and cerebellar vascular tumors

19. TUMOR MARKERS

β-HCG (Human Chorionic Gonadodotropin)

- gestational trophoblastic disease (e.g., choriocarcinoma, hydatidiform mole)
- dysgerminoma
- seminoma (10% of cases)

α-Fetoprotein (AFP)

- liver cancer
- germ cell tumors (e.g., yolk sac tumors, embryonal carcinoma, NOT seminoma)

Prostate-Specific Antigen (PSA) and Prostatic Acid Phosphatase (PAP)
• adenocarcinoma of prostate

Carcinoembryonic Antigen (CEA)
• adenocarcinomas of colon, pancreas, stomach, and breast (nonspecific marker)

CA-125
• ovarian cancer

S-100
• melanoma
• neural tumors

20. PROTEIN-ENERGY MALNUTRITION (PEM)

Kwashiorkor
• dietary protein deficiency (without calorie deficiency)
• anasarca (generalized edema)
• fatty liver (due to decreased apoproteins and decreased VLDL synthesis)
• abnormal skin and hair
• defective enzyme formation → malabsorption (hard to treat)

Marasmus
• dietary calorie deficiency (without protein deficiency)
• generalized wasting ("skin and bones")

21. NUTRITIONAL DEFICIENCIES

Vitamin A
• night blindness
• dry eyes and dry skin
• recurrent infections

Vitamin D
• decreased calcium
• bone → decreased calcification, increased osteoid
• children → rickets
• adults → osteomalacia

Vitamin E
• degeneration of posterior columns of spinal cord

Vitamin K
• decreased vitamin K–dependent factors → II, VII, IX, X, and proteins C and S
• increased bleeding
• increased PT and PTT

Vitamin B$_1$ (Thiamine)
• beriberi → wet (cardiac) or dry (neurologic)
• Wernicke-Korsakoff syndrome (lesions of mamillary bodies)

Vitamin B$_3$ (Niacin)
• pellagra → **3Ds** = **d**ermatitis, **d**ementia, **d**iarrhea (and death)

Vitamin B$_{12}$ (Cobalamin)
• megaloblastic (macrocytic) anemia
• hypersegmented neutrophils (> 5 lobes)
• subacute combined degeneration of the spinal cord

Vitamin C (Ascorbic Acid)
• scurvy
• defective collagen formation → poor wound healing (wounds reopen)
• bone → decreased osteoid
• perifollicular hemorrhages ("corkscrew" hair)
• bleeding gums and loose teeth

Folate
• megaloblastic (macrocytic) anemia
• hypersegmented neutrophils
• associated with neural tube defects in utero

Iron
• microcytic hypochromic anemia (with increased TIBC)

22. INHERITANCE PATTERNS

Autosomal Dominant (AD)
• disease produced in heterozygous state
• no skipped generations → parents affected (unless new mutation or reduced penetrance)

- father-to-son transmission possible
- males and females affected equally
- recurrence risk is 50%

Autosomal Recessive (AR)
- disease produced in homozygous state
- heterozygous individuals are carriers
- skipped generations
- father-to-son transmission possible
- males and females affected equally
- recurrence risk is 25%

X-Linked Dominant (XD)
- no skipped generations
- no male-to-male transmission
- females affected twice as often as males

X-Linked Recessive (XR)
- skipped generations
- no male-to-male transmission
- males affected more frequently than females

Y Inheritance
- only males affected
- only male-to-male transmission
- all males affected

Mitochondrial
- males and females affected
- only females transmit the disease

23. EXAMPLES OF XR
Hematology Diseases
- glucose-6-phosphate dehydrogenase (G6PD) deficiency
- hemophilia A (deficiency of factor VIII)
- hemophilia B (deficiency of factor IX)

Immunodeficiency Diseases
- Bruton's agammaglobulinemia
- chronic granulomatous disease
- Wiskott-Aldrich syndrome

Storage Diseases
• Fabry's disease
• Hunter's syndrome

Muscle Diseases
1. Duchenne muscular dystrophy
   • defective dystrophin gene (muscle breakdown)
   • pseudohypertrophy of calf muscles
   • Gower maneuver (using hands to rise from floor)
2. Becker muscular dystrophy

Metabolic Diseases
• diabetes insipidus
• Lesch-Nyhan syndrome

Other Diseases
• red-green color blindness
• fragile X syndrome

24. CHROMOSOMES

Terms
• haploid → number of chromosomes in germ cells (23)
• diploid → number of chromosomes found in nongerm cells (46)
• euploid → any exact multiple of the haploid number
• aneuploid → any nonmultiple of the haploid number
• triploid → three times the haploid number (69)
• tetraploid → four times the haploid number (92)
• trisomy → three copies of the same chromosome

25. AUTOSOMAL TRISOMIES

Trisomy 13 (Patau's Syndrome)
• mental retardation
• microcephaly and microphthalmia
• holoprosencephaly (fused forebrain)
• fused central face ("cyclops")
• cleft lip and palate
• heart defects

Trisomy 18 (Edwards' Syndrome)
- mental retardation
- micrognathia
- heart defects
- rocker-bottom feet
- clenched fist with overlapping fingers

Trisomy 21 (Down's Syndrome)
- most cases due to maternal nondisjunction during meiosis I (associated with increased maternal age)
- minority of cases due to Robertsonian (balanced) translocation
- mental retardation (most common familial cause)
- oblique palpebral fissures with epicanthal folds
- horizontal palmar crease
- heart defects (endocardial cushion defect is most common)
- acute lymphoblastic leukemia (first 2 years of life)
- Alzheimer's disease (almost 100% incidence after age 35)
- duodenal atresia ("double-bubble" sign on x-ray)

26. CHROMOSOMAL DELETIONS

5p– (Cri du Chat)
- high-pitched cry
- mental retardation
- heart defects and microcephaly

11p–
- Wilms tumor
- absence of iris

13q–
- retinoblastoma

15q–
1. Maternal deletion → Angelman's syndrome
   - stiff, ataxic gait with jerky movements
   - inappropriate laughter ("happy puppets")
   - may be due to two copies of paternal 15 chromosome (paternal uniparental disomy)

2. Paternal deletion → Prader-Willi syndrome
    - mental retardation
    - short stature and obesity
    - small hands and feet
    - hypogonadism
    - may be due to two copies of maternal 15 chromosome (paternal uniparental disomy)

27. HYPOGONADISM

Klinefelter's Syndrome
- most common genotype is 47,XXY
- male hypogonadism
- testicular dysgenesis → small, firm, atrophic testes
- decreased testosterone
- increased FSH, LH, estradiol
- decreased secondary male characteristics
- tallness, gynecomastia, and female distribution of hair
- infertility

Turner's Syndrome
- most common genotype is 45,XO
- female hypogonadism
- ovarian dysgenesis → streak ovaries
- decreased estrogen
- increased LH, FSH
- primary amenorrhea
- decreased secondary female characteristics
- skeletal abnormalities → short stature
- web neck (cystic hygroma)

28. AMBIGUOUS SEXUAL DEVELOPMENT

True Hermaphrodite
- ovaries and testes both present

Female Pseudohermaphrodite (XX Individual)
- ovaries
- male or ambiguous external genitalia
- due to excess androgens (e.g., congenital adrenal hyperplasia)

Male Pseudohermaphrodite (XY Individual)
- testes
- female external genitalia
- due to decreased androgen effects (most common → testicular feminization)

Androgen Insensitivity Syndrome (XY Individual)
- testicular feminization
- Müllerian duct regression (due to MIF)
- Wolffian duct regression (due to lack of testosterone receptors)
- phenotypic female (due to lack of receptors for DHT)

Decreased 5-α-Reductase (XY Individual)
- formation of testes (due to presence of Y chromosome)
- Müllerian duct regression (due to MIF)
- Wolffian duct development (due to testosterone)
- decreased DHT (due to lack of 5-α-reductase)
- variable external genitalia (due to decreased DHT)

Turner's Syndrome (XO Individual)
- streak gonads (due to lack of two X chromosomes)
- Müllerian duct development (due to lack of MIF)
- Wolffian duct regression (due to lack of testosterone)
- external female (due to lack of DHT)
- decreased secondary female characteristics (due to decreased estrogen)

Congenital Adrenal Hyperplasia (XX Individual)
- development of ovaries (due to two X chromosomes)
- Müllerian duct development (due to lack of MIF)
- Wolffian duct regression (due to lack of local testosterone production)
- external male (due to excess systemic formation of DHT)

29. DISORDERS OF TRINUCLEOTIDE REPEATS
1. Fragile X syndrome → CGG repeats
   - mental retardation (second most common familial cause; trisomy 21 is first)
   - long face with large ears
   - large testes (macroorchidism)
   - trinucleotide sequence expanded in females, not males

2. Huntington's syndrome → CAG repeats

3. Myotonic dystrophy → GCT repeats

4. Spinal-bulbar muscular atrophy → CAG repeats

30. LYMPHOCYTES

B Cells
- form plasma cells that secrete immunoglobulin
- surface antigen receptor composed of immunoglobulin
- rearrange immunoglobulin genes from germ line configuration
- CD19 → pan–B cell marker
- CD20 → pan–B cell marker, also called L26
- CD21 → pan–B cell marker, receptor for EBV
- CD22 → pan–B cell marker

T Cells
- secrete lymphokines
- surface antigen receptor (TCR) is attached to CD3
- rearrange genes for T cell receptor
- CD2 → receptor for sheep erythrocyte (E rosette)
- CD3 → attached to T cell receptor
- CD4 → helper T cells, bind with MHC class II antigens
- CD5 → pan–T cell marker
- CD7 → pan–T cell marker
- CD8 → cytotoxic T cells, bind with MHC class I antigens

Natural Killer Cells
- large granular lymphocytes
- do not need previous sensitization
- CD16 → receptor for Fc portion of IgG

31. IMMUNOGLOBULINS

IgM
- large molecule (pentamer)
- secreted early in immune response (primary response)
- cannot cross the placenta
- can activate complement
- contains a J chain

IgG
- most abundant immunoglobulin in serum
- secreted during second antigen exposure (secondary or amnestic response)
- can cross the placenta
- can activate complement
- can function as opsonin

IgE
- allergies, asthma, parasitic infection
- found attached to the surface of basophils and mast cells
- participates in type I hypersensitivity reactions

IgA
- usually a dimer with a J chain and a secretory component
- found along GI tract and respiratory tract
- secretory immunoglobulin
- can activate alternate complement pathway

IgD
- receptor for B cells
- found on the surface of mature B cells

32. T LYMPHOCYTES

CD4+ Cells
- helper T lymphocytes
- respond to MHC class II antigens

Subtypes:

1. T helper-1 (T_H1) cells
   - secrete → IL-2, IL-3, GM-CSF, γ-interferon, and lymphotoxin ($β$-TNF)
   - stimulate cell-mediated immune reactions → fight intracellular organisms
2. T helper-2 (T_H2) cells
   - secrete → IL-3, IL-4, IL-5, IL-6, IL-10, and GM-CSF
   - stimulate antibody production → fight extracellular organisms

CD8+ Cells

- cytotoxic T lymphocytes
- respond to MHC class I antigens

33. MAJOR HISTOCOMPATIBILITY COMPLEX (MHC)

Class I Antigens

- found on all nucleated cells
- transmembrane α glycoprotein chain with β_2-microglobulin
- react with antibodies and CD8-positive lymphocytes
- fight virus-infected cells and transplants

Class II Antigens

- found on antigen-presenting cells, B cells, and T cells
- transmembrane α chain and β chain
- react with CD4-positive lymphocytes
- fight exogenous antigens that have been processed by antigen-presenting cells

34. DISEASES ASSOCIATED WITH HLA TYPES

- ankylosing spondylitis → HLA-B27
- primary hemochromatosis → HLA-A3
- 21-hydroxylase deficiency → HLA-BW47
- rheumatoid arthritis → HLA-DR4
- insulin-dependent (type I) diabetes mellitus → HLA-DR3/DR4
- systemic lupus erythematosus → HLA-DR2/DR3

35. HYPERSENSITIVITY REACTIONS

Type I

- binding of antigen to previously formed IgE bound to mast cells and basophils
- release of histamine and leukotrienes C_4 and D_4
- urticaria (hives)
- anaphylaxis

Type II

- antibody (IgG or IgM) binds to antigens in situ
- cells destroyed by complement or cytotoxic cells (antibody-dependent cell-mediated cytotoxicity)

- linear immunofluorescence (IF)
- transfusion reactions

Type III
- antibody (IgG or IgM) binds to antigens forming immune complexes
- granular IF
- systemic → serum sickness
- local reaction → Arthus reaction

Type IV
1. Delayed type hypersensitivity
   - CD4 lymphocytes
   - extrinsic antigen associated with class II MHC
   - formation of activated macrophages (epitheloid cells) → granulomas
   - PPD skin test
   - contact dermatitis (poison ivy, poison oak)
2. Cell-mediated immunity
   - CD8 lymphocytes
   - intrinsic antigen associated with class I MHC
   - viral infections and transplant rejection

36. AUTOANTIBODIES
Nuclear
- diffuse (homogenous) → DNA (many diseases), histones (drug-induced SLE)
- rim (peripheral) → double-stranded DNA (SLE)
- speckled (non-DNA extractable nuclear proteins) → Smith (SLE), SS-A and SS-B (Sjögren's syndrome), Scl-70 (progressive systemic sclerosis)
- nucleolar (RNA) → many (e.g., progressive systemic sclerosis)
- centromere → CREST syndrome

Cytoplasmic
- mitochondria → primary biliary cirrhosis

Cells
- smooth muscle → lupoid hepatitis (autoimmune chronic active hepatitis)
- neutrophils → Wegener's granulomatosis and microscopic polyarteritis
- parietal cell and intrinsic factor → pernicious anemia
- microvasculature of muscle → dermatomyositis

Proteins
- immunoglobulin → rheumatoid arthritis
- thyroglobulin → Hashimoto's thyroiditis

Structural Antigens
- lung and glomerular basement membranes → Goodpasture's disease
- intercellular space of epidermis → pemphigus vulgaris
- epidermal basement membrane → bullous pemphigoid

Receptors
- acetylcholine receptor → myasthenia gravis
- thyroid hormone receptor → Graves' disease
- insulin receptor → diabetes mellitus

37. ANTINEUTROPHIL CYTOPLASMIC ANTIBODIES (ANCAs)

1. C-ANCAs (cytoplasmic)
   - proteinase 3 → Wegener's granulomatosis
2. P-ANCAs (perinuclear)
   - myeloperoxidase → microscopic polyarteritis

38. AMYLOIDOSIS

Amyloid
- any protein having β-pleated sheet tertiary configuration
- apple-green birefringence with Congo red stain

Systemic Deposition
- multiple myeloma → deposits of amyloid light protein
- chronic inflammatory diseases → deposits of amyloid-associated protein
- hemodialysis → deposits of β_2-microglobulin

Localized Deposition
- senile cardiac disease → deposits of amyloid transthyretin
- Alzheimer's disease → deposits of β_2-amyloid protein
- medullary carcinoma of thyroid → deposits of procalcitonin
- non-insulin-dependent diabetes mellitus (type II) → amyloid deposits in islets of Langerhans of pancreas

39. DEFECTS IN INFLAMMATION OR IMMUNITY

Chédiak-Higashi Syndrome
- autosomal recessive
- defective polymerization of microtubules
- giant lysosomes in leukocytes
- recurrent infections
- albinism (abnormal formation of melanin)

Chronic Granulomatous Disease
- defective NADPH oxidase (enzyme on membrane of lysosomes)
- recurrent infections with catalase-positive organisms
- abnormal nitroblue tetrazolium dye test

Severe Combined Immunodeficiency (SCID)
1. X-linked form
   - defect in IL-2 receptor
2. Autosomal recessive form (Swiss type)
   - lack of adenosine deaminase
   - prenatal diagnosis and gene therapy possible

X-Linked Agammaglobulinemia of Bruton
- defective maturation of B lymphocytes past the pre-B stage
- absence of germinal centers and plasma cells
- bacterial infections begin at the age of 9 months (loss of maternal antibody)
- therapy with immunoglobulin injections

Common Variable Immunodeficiency (CVID)
- variable clinical presentation
- recurrent infections → especially bacteria and *Giardia*
- hyperplastic B cell areas
- therapy with immunoglobulin injections

Isolated Deficiency of IgA
- most patients are asymptomatic
- may develop anti-IgA antibodies
- risk of anaphylaxis with transfusion

DiGeorge's Syndrome
• defective development of pharyngeal pouches 3 and 4
• deletion of chromosome 22
• lack of thymus → no T cells (recurrent viral and fungal infections)
• lack of parathyroid glands → hypocalcemia and tetany
• congenital heart defects

Acquired Immunodeficiency Syndrome (AIDS)
• cause → HIV infection
• infection of CD4+ T lymphocytes
• inversion of CD4/CD8 ratio (normal is 2:1)
• decreased humoral and cell-mediated immunity → recurrent infections
• increased incidence of malignancy (Kaposi's sarcoma and immunoblastic lymphoma)

40. VIRAL CHANGES

Giant Cells
• herpes simplex virus (HSV)
• cytomegalovirus (CMV)
• measles (Warthin-Finkeldey giant cells)
• respiratory syncytial virus

Inclusions
• herpes simplex virus (Cowdry A bodies)
• smallpox virus (Guarnieri bodies)
• rabies virus (Negri bodies)
• molluscum contagiosum (molluscum bodies)

Ground-Glass Change
• nucleus = herpes simplex virus
• cytoplasm (of hepatocytes) = hepatitis B

Atypical Cells
• atypical lymphocytes → Epstein-Barr virus
• smudge cells → adenovirus (respiratory epithelial cells)
• koilocytosis → human papillomavirus (HPV)

41. SYSTEMIC MYCOSES

Candidiasis
• *Candida albicans*
• pseudohyphae
• white plaques (thrush)

Histoplasmosis
- *Histoplasma capsulatum*
- found within the cytoplasm of macrophages
- bird droppings; bat guano in caves
- Ohio and Mississippi valleys

Aspergillosis
- *Aspergillus* species
- septate hyphae with acute-angle branching
- fruiting bodies (when exposed to air → fungus ball in lung cavity)

Blastomycosis
- *Blastomyces dermatitidis*
- broad-based budding
- eastern United States

Coccidioidomyocosis
- *Coccidioides immitis*
- large spherules filled with many small endospores
- southwestern United States (San Joaquin Valley)

Cryptococcosis
- *Cryptococcus neoformans*
- CNS infection in immunosuppressed patients
- mucicarmine-positive capsule
- india ink stain of CSF

Mucormycosis
- nasal infection in diabetic patients
- broad, nonseptate hyphae with right-angle branching

42. FAMILIAL HYPERLIPIDEMIA

Type I Hyperlipoproteinemia
- familial hyperchylomicronemia
- mutation in lipoprotein lipase gene
- increased serum chylomicrons

Type II Hyperlipoproteinemia
- familial hypercholesterolemia
- mutation involving LDL receptor
- increased serum LDL
- increased serum cholesterol

Type III Hyperlipidemia
- floating or broad β disease
- mutation in apolipoprotein E
- increased chylomicron remnants and IDL
- increased serum triglycerides and cholesterol

Type IV Hyperlipidemia
- familial hypertriglyceridemia
- unknown mutation
- increased serum VLDL
- increased serum triglycerides and cholesterol

Type V Hyperlipidemia
- mutation in apolipoprotein CII
- increased serum chylomicrons and VLDL
- increased serum triglycerides and cholesterol

43. ANEURYSMS
Atherosclerotic Aneurysms
- cause → atherosclerosis
- location → abdominal aorta (between renal arteries and bifurcation of the aorta)
- pulsatile mass
- may rupture → sudden, severe abdominal pain in male older than 55
- treat with surgery when diameter is > 5 cm

Luetic Aneurysms
- cause → syphilis (treponema) infection
- obliterative endarteritis (plasma cells around small blood vessels)
- location → ascending (thoracic) aorta
- may produce aortic regurgitation or rupture

Dissecting Aneurysms
1. Due to cystic medial necrosis of aorta
   - hypertension
   - Marfan's syndrome → due to defect in fibrillin gene
2. "Double-barrel" aorta on x-ray

Berry Aneurysms
- location → bifurcation of arteries in circle of Willis
- most commonly bifurcation of anterior communicating artery
- subarachnoid hemorrhage
- associated with polycystic renal disease

44. CARDIAC HYPERTROPHY

Concentric Hypertrophy
- response to pressure overload (e.g., hypertension or aortic stenosis)
- sarcomeres proliferate in parallel
- increased ventricular thickness
- no change in size of ventricular cavity

Eccentric Hypertrophy
- response to volume overload
- sarcomeres proliferate in series
- no increase in ventricle thickness
- increase in size of ventricular cavity

45. CONGENITAL HEART DEFECTS

Left-to-Right Shunts
1. Ventricular septal defect (VSD) → most common congenital cardiac anomaly
2. Atrial septal defect (ASD)
3. Patent ductus arteriosus (PDA)
   - "machine-like" heart murmur
   - indomethacin closes PDA

Right-to-Left Shunts
1. Tetralogy of Fallot (TOF) → most common cause of congenital cyanotic heart disease
   - pulmonary stenosis
   - ventricular septal defect
   - dextropositioned (overriding) aorta
   - right ventricular hypertrophy

No Shunts

1. Coarctation of the aorta
   • infantile type (preductal)
   • adult type (postductal) → rib notching, increased BP in upper extremities, decreased BP in lower extremities
2. Transposition of the great vessels
   • need shunt to be present in order to survive (e.g., PDA)
   • PGE keeps ductus open

46. ATROPHY OF THE STOMACH

Type A → Autoimmune Gastritis
• autoantibodies to parietal cells and intrinsic factor → pernicious anemia
• decreased vitamin B_{12} → megaloblastic anemia
• increased serum gastrin levels
• histologic changes found in fundus of stomach

Type B → Environmental
• no autoantibodies present
• associated with *Helicobacter pylori* (urease breath test is positive)
• decreased serum gastrin levels
• histologic changes found in antrum of stomach

47. INFLAMMATORY BOWEL DISEASE (IBD)

Ulcerative Colitis
• crypt abscesses (microabscesses) and crypt distortion
• disease begins in rectum and extends proximally (no skip lesions)
• does not involve small intestines
• superficial mucosal involvement (not transmural)
• increased risk of colon cancer and toxic megacolon

Crohn's Disease
• granulomas
• segmental involvement (skip lesions)
• may involve small intestines (regional enteritis or ileitis)
• transmural involvement → fissures, fistulas, and obstruction

48. GALLSTONES

Cholesterol Stones
- yellow stones
- risk factors → **Fs** = fat, female, fertile, forty, fifty
- increased incidence in Native Americans

Bilirubin (Pigment) Stones
- black stones
- risk factors → chronic hemolysis and infections of biliary tract
- increased incidence in Asians

49. CONGENITAL ADRENAL HYPERPLASIA (CAH)

21-Hydroxylase Deficiency
- decreased cortisol → increased ACTH
- decreased aldosterone
- sodium loss in the urine → salt-wasting form of CAH
- hyperkalemic acidosis
- virilism in females

11-Hydroxylase Deficiency
- decreased cortisol → increased ACTH
- decreased aldosterone
- increased DOC and 11-deoxycortisol → increased mineralocorticoid effects
- sodium retention → hypertensive form of CAH
- hypokalemic alkalosis
- virilism in females

17-Hydroxylase Deficiency
- decreased cortisol → increased ACTH
- no decreased aldosterone
- decreased sex hormones
- females → primary amenorrhea
- males → pseudohermaphrodites

50. MULTIPLE ENDOCRINE NEOPLASIA

Type 1 (Wermer's Syndrome)
- parathyroid
- pituitary
- pancreas

Type 2 (Sipple's Syndrome)
- parathyroid
- medullary carcinoma of thyroid
- pheochromocytoma

Type 3 (MEN 2B)
- medullary carcinoma of thyroid
- pheochromocytoma
- mucosal neuromas

51. RENAL (GLOMERULAR) SYNDROMES

Nephrotic Syndrome
- marked proteinuria → hypoalbuminemia and edema
- increased cholesterol → oval fat bodies in the urine

Examples (nonproliferative glomerular disease):

1. Minimal change disease (lipoid nephrosis)
   - normal light microscopy
   - EM reveals fusion of foot processes of podocytes
2. Focal segmental glomerulosclerosis (FSGS)
3. Membranous glomerulonephropathy (MGN)
   - thickening of basement membrane ("spikes and domes")
   - uniform subepithelial deposits
4. Diabetes mellitus

Nephritic Syndrome
- hematuria (red blood cells and red blood cell casts in urine)
- variable proteinuria and oliguria
- retention of salt and water (hypertension and edema)

Examples (proliferative glomerular disease):

1. Focal segmental glomerulonephritis (FSGN)
   - mesangial deposits of IgA
   - Berger's disease

2. Acute (diffuse) proliferative glomerulonephritis (DPGN)
   - post-streptococcal glomerulonephritis
   - large, irregular subepithelial deposits

3. Membranoproliferative glomerulonephritis (MPGN)
   - subendothelial deposits → type I MPGN
   - intramembranous deposits → type II MPGN (dense deposit disease)
   - splitting of basement membrane by mesangium → "tram-track" appearance

4. Rapidly progressive glomerulonephritis (RPGN)

52. GLOMERULAR DEPOSITS

Subepithelial
- diffuse proliferative glomerulonephritis (DPGN) → irregular and large
- membranous glomerulonephropathy (MGN) → uniform and small

Intramembranous (Basement Membrane)
- membranoproliferative glomerulonephritis (MPGN), type II

Subendothelial
- membranoproliferative glomerulonephritis, type I
- SLE

Mesangial
- focal segmental glomerulonephritis (FSGN)
- Henoch-Schönlein purpura

53. RAPIDLY PROGRESSIVE GLOMERULONEPHRITIS (RPGN)

Linear Immunofluorescence
- antimembrane antibody
- Goodpasture's disease

Granular Immunofluorescence
- immune complexes
- other glomerular or systemic disease

Minimal or Negative Immunofluorescence
- pauci-immune disease
- Wegener's granulomatosis
- microscopic polyarteritis nodosa

54. CEREBRAL HEMORRHAGE

Epidural Hematoma
- severe trauma
- arterial bleeding (middle meningeal artery)
- symptoms occur rapidly

Subdural Hematoma
- minimal trauma in elderly
- venous bleeding (bridging veins)
- symptoms occur slowly

Subarachnoid Hemorrhage
- rupture of berry aneurysm
- "worst headache ever"
- bloody or xanthochromic spinal tap

55. INFECTIONS OF THE MENINGES

Bacterial Infections
- increased neutrophils and protein in CSF
- decreased glucose in CSF
- life-threatening

Age	Organism
Neonates	*Escherichia coli*
6 months to 6 years	*Streptococcus pneumoniae*
6 years to 16 years	*Neisseria meningitidis* (meningococcus)
Older than 16 years	*Streptococcus pneumoniae*
Epidemics	*Neisseria meningitidis*

Viral Infections
- increased lymphocytes in CSF
- normal glucose in CSF
- mild and self-limited

56. ATROPHY OF THE NERVOUS SYSTEM

Alzheimer's Disease
- diffuse atrophy of cerebral cortex
- dementia (most common cause in elderly)
- senile plaques (with β-amyloid core)
- neurofibrillary tangles (with abnormal τ protein)

Pick's Disease
- unilateral frontal or temporal lobe atrophy

Huntington's Disease
- trinucleotide repeat disorder
- atrophy of caudate and putamen → decreased GABA and acetyl-choline
- progressive dementia
- choreiform movements

Parkinson's Disease
- substantia nigra (depigmentation)
- decreased dopamine in corpus striatum
- cogwheel rigidity and akinesia
- tremor
- treatment → dopamine agonists

57. JOINTS

Rheumatoid Arthritis
- rheumatoid factor (IgM antibody against antibody)
- pannus formation in synovium (hyperplastic synovium with lymphocytes and plasma cells)
- ulnar deviation of fingers
- subcutaneous rheumatoid nodules (at pressure points)
- pain worse in morning ("morning stiffness"); pain decreases with activity

Osteoarthritis

- degenerative joint disease ("wear and tear")
- loss of articular cartilage → smooth subchondral bone (eburnation)
- osteophyte formation (DIP → Heberden's nodes, PIP → Bouchard's nodes)
- pain worse in evening; pain increases with activity

Gout

- hyperuricemia → precipitation of monosodium urate crystals (needle-shaped, negatively birefringent crystals)
- first MTP joint (big toe)
- tophus formation
- increased production of uric acid → Lesch-Nyhan syndrome
- increased turnover of nucleic acid → leukemias and lymphomas
- decreased excretion of uric acid → chronic renal disease, ethanol intake, diabetes

58. STAINS

Routine (H&E)

1. Hematoxylin
   - blue and basic
   - stains negatively charged structures → DNA and RNA
2. Eosin
   - pink and acidophilic
   - stains positively charged structures → mitochondria

Special Stains

- fats → oil red O
- glycogen → PAS-positive, diastase-sensitive
- iron → Prussian blue
- hemosiderin → Prussian blue
- amyloid → Congo red
- α_1 antitrypsin → PAS-positive, diastase-resistant
- calcium → von Kossa

59. ENZYMES

Aminotransferases (AST, ALT)

- myocardial infarction (AST)

- alcoholic hepatitis (AST > ALT)
- viral hepatitis (ALT > AST)

Creatine Kinase (CK or CPK)
- myocardial infarction (CPK-MB)
- muscle diseases (DMD)

Lactate Dehydrogenase (LDH)
- myocardial infarction (LDH1, LDH2)

Amylase or Lipase
- acute pancreatitis

60. HISTOLOGIC "BODIES"

1. Psammoma body:
   - papillary carcinoma of the thyroid
   - papillary tumors of the ovary
   - meningioma

2. Immunoglobulin
   - Russell body → cytoplasmic or extracellular
   - Dutcher body → nucleus (Waldenstrom's)

3. Councilman body → viral hepatitis

4. Mallory body → alcoholic hyaline

5. Cowdry A body → herpes

6. Aschoff body → rheumatoid fever

7. Ferruginous body → asbestos

8. Negri body → rabies

9. Lewy body → Parkinson's

10. Heinz body (denatured hemoglobin) → G6PD deficiency

11. Barr body → number of X chromosomes minus one

61. HEALING OF THE MYOCARDIUM AFTER A MYOCARDIAL INFARCTION

	Gross	Microscopy
0–12 h	None	Usually none (?wavy fibers)
12–24 h	Pallor	Coagulative necrosis
1–3 days	Hyperemic (red) border	Above + neutrophils
4–7 days	Pale yellow	Above + macrophages
7–14 days	Red-purple border	Above + granulation tissue
>2 weeks	Gray-white scar	Fibrosis (scar)

62. FAMILIAL STORAGE DISORDERS

Storage Disease	Enzyme Deficiency	Substance Accumulating
Pompe's disease	α-1,4-glucosidase (acid maltase)	Glycogen
Hurler's syndrome	α-L-iduronidase	Heparan sulfate, dermatan sulfate
Hunter's syndrome	sc-l-iduronosulfate sulfatase	Heparan sulfate, dermatan sulfate
Niemann-Pick disease	Sphingomyelinase	Sphingomyelin
Tay-Sachs disease	Hexosaminidase A	G_{M2} ganglioside
Sandhoff's disease	Hexosaminidase A and B	G_{M2} ganglioside and globoside
Gaucher's disease	Glucocerebrosidase	Glucocerebroside
Fabry's disease	α-galactosidase A	Ceramide trihexosidase

High-Yield Facts in Pathophysiology

1. Many diseases have an immunologic basis. Example: **Graft versus host (GVH) disease** can develop in an immunosuppressed individual who receives immunocompetent donor cells. The donor cells respond to histocompatibility antigens present on the recipient's cells that are NOT found on the donor cells. Bone marrow contains immunocompetent T cells.

2. The following chart compares **bacterial meningitis** and **viral meningitis.**

	Bacterial Meningitis	Viral Meningitis
Disease state	Acute: significant mortality without antibiotic therapy	Acute: usually self-limited
Symptoms	Fever	Fever
	Worst headache of life	Worst headache of life
	Meningismus	Meningismus
	Mental status changes	Mental status changes
Physical exam findings	Photophobia	Photophobia
	Nausea	Nausea
	Vomiting	Vomiting
	Fever	Fever
	Kernig's sign—positive	Kernig's sign—positive
	Brudinski's sign—positive	Brudinski's sign—positive
Etiology	Neonates	Cocksackie A and B viruses
	Escherichia coli	Poliovirus
	Group B *Streptococcus*	Mumps virus
	Listeria monocytogenes	Epstein-Barr virus
	Children	Adenovirus
	Neisseria meningitidis	Cytomegalovirus
	Streptococcus pneumoniae	
	Haeophilus influenzae, nonimmunized	

Continued

	Bacterial Meningitis	**Viral Meningitis**
Etiology (*cont'd*)	Adults (more than 18 years old) *N. meningitidis* *S. pneumoniae* *L. monocytogenes* Gram-negative bacilli	
Cerebrospinal fluid results	Decreased glucose Increased protein Increased neutrophils Increased pressure	Normal glucose Slightly increased protein Increased monocytes Normal or slightly increased pressure Gram stain shows no bacteria
Treatment	IV antibiotics Supportive therapy	Supportive therapy
Complications	Cerebral edema Deafness Death	Deafness Weakness

3. Carcinomas undergo phenotypic transition from **normal → hyperplasia → carcinoma in situ → invasive carcinoma → metastasis**. Carcinomas occur as a result of a constellation of physiologic and genetic changes (e.g., APC, hMLH1, and hMSH2—colon carcinoma/BRCA1 and BRCA2—breast carcinoma).

4. **Colon carcinoma** begins when cell cycle regulation loses control over growth, and a collection of rapidly multiplying cells (**hyperplasia**) form an adenoma. The adenoma can continue to develop into **carcinoma in situ**. The first evidence of disease may be occult rectal bleeding indicating the appearance of new friable vessels supplying the tumor. Next, the cancer cells invade the basement membrane of the colon (**invasive carcinoma**), gaining access to the body's transport systems (lymphatic and hematogenous). **Metastasis** to lymph nodes and distant body regions can occur.

5. Many malignancies have characteristic indirect systemic effects via multiple mechanisms. In lung malignancies, excess adrenocorticotropic hormone (ACTH) production results in a Cushing-like syn-

drome and excess antidiuretic hormone (ADH) production results in a syndrome of inappropriate antidiuretic hormone secretion (SIADH). Malignancies (e.g., squamous cell carcinoma) can produce peptides related to PTH, causing hypercalcemia. Carcinoid syndromes produce serotonin or prostaglandins that can cause flushing, restrictive lung symptoms, ascites, and hypotension.

6. **Pernicious anemia** occurs when antibodies to intrinsic factor and parietal cells attack the gastric mucosa, causing gastric atrophy. The disruption of the normal function of the gastric mucosa affects vitamin B_{12} absorption on two levels: stomach acid deficiency (achlorhydria) prevents the release of vitamin B_{12} from food digestion, and intrinsic factor is necessary for vitamin B_{12} absorption in the terminal ileum. The chronic loss of vitamin B_{12} results in abnormal RBC maturation without changes in hemoglobin synthesis leading to macrocytic anemia.

7. Pathophysiology of hearing loss.

Type of Hearing Loss	Etiology	Testing
Conductive deafness	Disruption of conduction and amplification of sound from the external auditory canal to the inner ear	Negative Rinne test Weber test: heard best in the affected ear Audiometry
Sensorineural	Impaired function of inner ear or cranial nerve VIII	Positive Rinne test Weber test: heard best in the unaffected ear Audiometry
Central deafness	Damaged CNS auditory pathways	Audiometry

8. **Myasthenia gravis** is an autoimmune disease characterized by antibodies to acetylcholine receptors, causing a deficiency in the number of acetylcholine receptors on the postsynaptic (muscle) terminal, resulting in reduced efficiency of neuromuscular activity. The disease commonly presents in small muscle groups, accompanied by intermittent fatigue and weakness relieved by rest.

9. **Psoriasis** is an inflammatory parakeratotic accumulation of skin cells that features erythematous, demarcated lesions with scaly patches commonly found on scalp, extensor surfaces of extremities, and fingernails.

10. **Asthma** is an obstructive pulmonary disease characterized by airway narrowing as a result of smooth muscle spasms, inflammation, edema, and thick mucus production. The pathophysiologic response is mediated by local cellular injury, lymphocyte activation (antigen exposure, B cell activation, and cytokine activity), IgE-mediated mast cell (producing histamine, leukotrienes, and platelet-activating factor), and eosinophil activation.

11. Pulmonary function tests: obstructive lung disease vs. restrictive lung disease.

Pulmonary Function Test	Obstructive Lung Diease (e.g., Chronic Obstructive Pulmonary Disease)	Restrictive Lung Disease (e.g., Pulmonary Fibrosis)
FVC	↓	↓
FEV$_1$	↓	↓
FEV$_1$%	↓	Normal / ↑
TLC	↑	↓
RV	↑	Normal / ↓

12. **Pulmonary embolism** occurs when a venous thrombi (usually from a deep vein thrombosis) lodges in the pulmonary circulation. The pathophysiology includes hemodynamic changes, increased alveolar dead space with increased ventilation/perfusion ratios, and decreased oxygen perfusion to body tissues. Common acute presentations include tachypnea, hemoptysis, tachycardia, fever, cough, and pleuritic pain.

13. In normal individuals, as left ventricular end-diastolic pressure or preload increases, stroke volume will increase proportionately. In patients who suffer heart failure, increased left ventricular end-diastolic pressure is not met with increased stroke volume, because the contractility is depressed and is unable to function; thus, the patient ultimately experiences heart failure. Frank-Starling curves or ventricular function curves are diagrams that show the relationship between stroke volume or cardiac output and preload or left ventricular end-diastolic volume.

14. **Stable angina** is caused by a fixed partial atherosclerotic plaque in one or more of coronary arteries. When at rest, blood flow is able to provide adequate oxygenation to the heart muscle. On exertion, oxygen demand increases. The partial occlusion prevents adequate oxygenation to the heart, resulting in chest discomfort. Unstable angina is caused by thrombus formation on a fissuring atherosclerotic plaque, which transiently prevents adequate oxygenation to the heart. The resulting ischemia causes chest discomfort whether at rest or during exertion.

15. Chronic esophageal reflux (as a result of a transient weakened lower esophageal sphincter, alcohol use, and tobacco abuse) can result in **Barrett's esophagus.** In the disease, columnar epithelium replaces normal squamous epithelium. Individuals with Barrett's esophagus have an increased risk of developing adenocarcinoma of the esophagus.

16. **Helicobacter pylori** is a common bacteria that infects the gastric mucosa, providing an increased propensity for peptic ulcer disease through inflammatory mechanisms. Other risk factors for peptic ulcer disease are use of a nonsteroidal anti-inflammatory drug (NSAID), family history, smoking, and Zollinger-Ellison syndrome (gastrinoma).

17. **Crohn's disease** is a chronic inflammatory bowel disease that affects the whole gastrointestinal tract (from mouth to anus) and is distinguished by **alternating regions** ("skip lesions") of normal bowel and full-thickness ulcerations and granuloma formation of the bowel wall. Common manifestations are bloody diarrhea, fistula, iritis, arthritis, abscess formation, and small bowel obstruction.

18. **Ulcerative colitis** is an inflammatory bowel disease that causes **continuous,** partial-thickness (mucosa only) ulcerations of all or part of the colon and is manifested by bloody diarrhea and abdominal pain.

19. **Type 1 diabetes mellitus** (previously called insulin-dependent diabetes mellitus) and **type 2 diabetes mellitus** (previously called non-insulin-dependent diabetes) differ in multiple ways. Type 1 DM usually starts in young (less than 30 years old), nonobese individuals who sometimes have a family history (weak genetic component). Insulin production deficiency predominates, with rare insulin receptor resistance. Type 1 DM is always treated with exogenous insulin. Diabetic ketoacidosis is a common complication. Type 2 DM usually starts in older (more than 40 years old), obese individuals who often have a family history of disease

(strong genetic component). Recently, type 2 DM has been seen in younger adults and children, probably due to a change in life style. Insulin resistance is a major feature. Type 2 DM is often treated with exogenous insulin. Hyperosmolar coma is a common complication.

20. **Hyperthyroidism** is characterized by sweating, agitation, weight loss, heat intolerance, palpitations, irritability, and dyspnea. Triiodothyronine (T_3) and thyroxine (T_4) are elevated with concurrent depression of thyroid-stimulating hormone (TSH). Hypothyroidism is characterized by fatigue, depression, constipation, weight gain, decreased sweating, cold intolerance, and hoarseness. T_3 and T_4 are depressed with concurrent elevation of TSH.

21. **Cushing's syndrome** (excessive cortisol production) is characterized by moon facies, neck/trunk obesity, weight gain, mental status changes, purple striae on abdomen, osteoporosis, and glucose intolerance. Often there is excess ACTH production from the pituitary (Cushing's disease) or from tumors (ectopic ACTH).

22. **Conn's syndrome** (excessive mineralocorticoid secretion) is characterized by hypokalemia, hypertension, metabolic alkalosis, glucose intolerance, and weakness.

23. **Addison's disease** (primary adrenal insufficiency) is characterized by weakness, fatigue, weight loss, hypotension, cold intolerance, abdominal pain, diarrhea, anorexia, hyperkalemia, hyponatremia, and hypoglycemia.

24. **Preeclampsia-eclampsia** is characterized by hypertension, proteinuria, and edema after week 20 of pregnancy. Without treatment, a pattern of complications occurs. Complications include bleeding, malignant hypertension, stroke, renal failure, seizures, disseminated intravascular coagulation, and death.

25. **Minimal change disease** is the most common cause of nephrotic syndrome in children and is characterized by isolated proteinuria (more than 3.5 g of protein in 24-h urine) and obliterated epithelial podocytes on the glomerular basement membrane.

26. **Human immunodeficiency infection (HIV) and acquired immunodeficiency disease (AIDS).** HIV infection occurs worldwide. It is mainly a sexually transmitted disease that can be transmitted by blood and blood products contaminated with the virus. The diagnosis of

AIDS is made when any AIDS defining illness occurs in a person with HIV, such as **Pneumocystis carinii pneumonia (PCP)**, thrush due to *Candida*, or their CD4 lymphocyte count drops below 200 cells/μL. AIDS is the leading cause of death in persons 25–44 years old in the United States. HIV infection begins with an acute HIV syndrome, latency ensues, and then early symptomatic illness and eventually fully symptomatic illness with a myriad of complicating infections and non-infectious diseases. The entire cycle occurs over a period of more than 10–15 years, and possibly longer now that effective treatment regimens are available. The treatment of HIV and AIDS can alter this progression and provide long periods free of detectable viral loads, almost normal CD4 lymphocyte counts and good quality of life. Treatment changes rapidly on the basis of the introduction of new anti-HIV drugs and new protocols specifying their use. The standard of care now is three-drug regimens (triple therapy).

High-Yield Facts
in Pharmacology

Sample Drug Classification Tables

TIPS FOR LEARNING PHARMACOLOGY

Pharmacology is best learned by comparing drugs within a particular class or by their specific use.

A chart highlighting the similarities and differences among the various agents can be a helpful tool. The charts included in this section are simple examples. More elaborate charts can be constructed that would include how the drug is administered, its pharmacological effects, its adverse effects, its mechanism of toxicity (if known), and significant drug-drug interactions. For infectious disease agents, the spectrum of antimicrobial activity and the basis of antibiotic resistance can be added.

Explanations for the abbreviations used in these charts are found in the List of Abbreviations and Acronyms, which appears before the Bibliography.

Drugs for Treating Bacterial Infectious Diseases

Drug Class	Prototype	Action	Spectrum
Penicillins		Inhibit bacterial cell-wall synthesis by binding to penicillin-binding proteins, inhibiting crosslinking enzymes, and activating autolytic enzymes that disrupt bacterial cell walls.	Streptococci, meningococci, pneumococci, gram-positive bacilli, gonococci, spirochetes.
Narrow spectrum			
Penicillinase-susceptible	Penicillin G		
Penicillinase-resistant	Methicillin		Staphylococci.
Wide spectrum	Ampicillin		Similar to penicillin G; also includes *E. coli, P. mirabilis,* and *H. influenzae.*
Penicillinase-susceptible	Carbenicillin		Gram-negative rods and especially useful for *Pseudomonas* spp.
Cephalosporins			
First-generation	Cephalothin		Gram-positive cocci, *E. coli,* and *K. pneumoniae.*
Second-generation	Cefamandole		Greater activity against gram-negative organisms than first-generation cephalosporins.
Third-generation	Cefoperazone		Broader activity against resistant gram-negative organisms; some derivatives penetrate the blood-brain barrier.
Carbapenem	Imipenem		Wide action against gram-positive cocci, gram-negative rods, and some anaerobes.
Monobactam	Aztreonam		Resistant to β-lactamases produced by gram-negative rods.

Macrolides	Erythromycin	Inhibits protein synthesis by binding to part of the 50S ribosomal subunit	Gram-positive cocci, mycoplasma, corynebacteria, *Legionella, Ureaplasma, Bordetella.*
Vancomycin	Vancomycin	Inhibits synthesis of cell-wall mucopeptides (peptidoglycans).	Gram-positive bacteria, especially for resistant mutants.
Chloramphenicol	Chloramphenicol	Inhibits peptide bond formation by binding to the 50S ribosomal subunit, inhibiting peptidyl transferase.	*Salmonella* and *Haemophilus* infections and meningococcal and pneumococcal meningitis.
Aminoglycosides Systemic	Gentamicin	Inhibits protein synthesis by binding to the 30S subunit of ribosomes, which blocks formation of the initiation complex, causing misreading of the code on the mRNA template and disrupting polysomes.	*E. coli, Enterobacter, Klebsiella, Proteus, Pseudomonas,* and *Serratia* species.
Local	Neomycin		
Tetracycline	Tetracycline	Inhibits protein synthesis by binding to the 30S ribosomal subunit, which interferes with binding of aminoacyl-tRNA.	Mycoplasma, chlamydia, rickettsia, vibrio.
Sulfa drugs	Sulfonamides	Inhibit folic acid synthesis by competitive inhibition of dihydropteroate synthase.	Gram-positive and -negative organisms, including chlamydia and nocardia.
Trimethoprim	Trimethoprim	Inhibits folic acid synthesis by inhibition of dihydrofolate reductase.	Used in combination with sulfamethoxazole.
Fluoroquinolones	Norfloxacin	Inhibits topoisomerase II (DNA gyrase).	Gram-negative organisms, including gonococci; *E. coli, K. pneumoniae, C. jejuni, Enterobacter, Salmonella,* and *Shigella* species.

Drugs for Treating Hypertension		
Drug Class	**Prototype**	**Action**
Sympathetic nervous system agents		
Central	Clonidine	α_2-agonist; causes decreased sympathetic outflow.
Peripheral	Guanethidine	Uptake by transmitter vesicles in nerve depletes and replaces norepinephrine in neurosecretory vesicles.
	Prazocin	α_1-antagonist.
	Propranolol	β-antagonist.
Central and peripheral	Reserpine	Binds tightly to storage vesicles, which consequently lose their ability to concentrate and store norepinephrine.
Vasodilators		
Arterial	Hydralazine	Unknown.
	Diazoxide	Opens K^+ channels and causes hyperpolarization of smooth muscle.
Arterial and venous	Nitroprusside	Releases NO, which binds to guanylyl cyclase to generate cGMP.
Ca^{++} channel-blockers	Nifedipine	Inhibits voltage-dependent "L-type" Ca^{++} channels.
ACE inhibitors	Captopril	Inhibits conversion of angiotensin I to angiotensin II.
Diuretics		
Thiazides (benzothiadiazides)	Hydrochlorothiazide	Inhibits Na^+ channels in luminal membrane in the proximal segment of the distal tubule.
Loop agents	Furosemide	Inhibits cotransporter of Na^+, K^+, Cl^- in the ascending limb of the loop of Henle.

High-Yield Facts

General Principles

Serum concentration vs time
  graphs
Relationship of drug elimination
  half-time ($t_{1/2}$)
Apparent volume of distribution
Drug clearance
Drug distribution
  Henderson-Hasselbalch
    equations
  Diffusion
Partition coefficients
Bioavailability
Log-dose response curves

Anti-Infectives

Cell-wall synthesis inhibitors
  Penicillins
  Cephalosporins
  Monobactams
  Carbapenem
  Vancomycin
  Cycloserine
β-lactamase inhibitors
Protein synthesis inhibitors
  Chloramphenicol
  Tetracyclines
  Macrolides
  Lincosamides
  Aminoglycosides

Folic acid synthesis inhibitors
  Sulfonamides
  Trimethoprim
DNA synthesis inhibitors
  Fluoroquinolones
Antimycobacterials
  Isoniazid
  Rifampin
  Ethambutol
  Pyrizinamide
  Streptomycin
Antileprosy agents
Antifungals
  Amphotericin B
  Flucytosine
  Azoles
  Terbinafine
Antivirals
  Antiherpes agents
Antiretrovirals
  Nucleoside reverse transcriptase
    inhibitors
  Nonnucleoside reverse
    transcriptase inhibitors
  Protease inhibitors
  Amantadine
  Interferons
  Ribavirin
Antiprotozoals
Antihelminthics

Organism	Drug
Pneumococcus	Penicillin G, ampicillin
Pneumococcus (penicillin-resistant)	Fluoroquinolones
Streptococcus	Penicillin G, macrolides (allergic patients)
Staphylococcus (penicillinase-resistant)	Penicillinase-resistant penicillin
Staphylococcus (methicillin-resistant)	Vancomycin
Enterococcus	Penicillin G and gentamycin
Enterococcus (vancomycin-resistant)	Linezolid
Gonococcus	Ceftriaxone, fluoroquinolones
Menigicoccus	Penicillin G, ampicillin, cephtriaxone
Escherichia coli, Proteus, Klebsiella	Second- and third-generation cephalosporin, trimethoprim-sulfamethoxazole, ampicillin, fluoroquinolones
Shigella	Fluoroquinolones
Enterobacter, Serratia	Imipenem, trimethoprim-sulfamethoxazole, fluoroquinolones, pipericillin/tazobactam
Hemophilus	Second- or third-generation cephalosporins, trimethoprim-sulfamethoxazole, fluoroquinolones
Pseudomonas	Cephtazidime, cefepime, imipenem, aztreonam, ciprofloxacin, aminoglycoside, and extended-spectrum penicillin
Bacteroides	Metronidazole, clindamycin
Mycoplasma	Macrolide, tetracycline
Treponema	Penicillin G

Drug	Adverse Drug Reaction
Penicillins	Cross-allergenicity
Cephalosporins	Cross-allergenicity
	Contraindicated in patients with history of anaphylaxis to penicillins
	Disulfiram-like reaction with ethanol
Vancomycin	"Red person" syndrome
Chloramphenicol	"Gray baby syndrome," aplastic anemia
Macrolides	Arrhythmias with coadministration of astemizole

Drug	Adverse Drug Reaction
Clindamycin	Clostridium difficile colitis
Aminoglycosides	Ototoxicity and nephrotoxicity
Tetracycline	Discolored teeth, enamel dysplasia, and bone growth disturbances in children
Sulfa drugs	Cross-allergenicity with other sulfa drugs and with certain diuretics and hypoglycemics
Fluoroquinolones	Tendonitis, Achilles tendon rupture, contraindicated in patients less than 18 years old because of effects on cartilage development
Amphotericin B	Shocklike reaction
Azole antifungals	Arrhythmias with astemizole
Isoniazid	Hepatotoxicity prevented by coadministration of pyridoxine
Ethambutol	Visual disturbances
Pyrazinamide	Nongouty polyarthralgias
Dapsone	Hemolysis in patients with glucose-6-phosphate dehydrogenase deficiency

Antiviral Agent	Adverse Drug Reaction
Zidovudine (AZT)	Anemia
Didanosine (ddI)	Neuropathy, pancreatitis
Stavudine (d4T)	Neuropathy
Abacavir	Hypersensitivity reaction
Efavirenz	Central nervous system toxicity
Protease inhibitors	Hepatotoxicity, hyperlipidemia, nephrolithiasis, lipodystrophy
Acyclovir	Nephropathy
Ganciclovir	Neutropenia
Foscarnet	Renal toxicity
Ribavirin	Anemia
Interferons	Flulike symptoms
Lamivudine	Lactic acidosis
Rimantadine, amantadine	Central nervous system toxicity
Zanamavir	Bronchospasm

Cancer Chemotherapy and Immunology

Cell cycle kinetics
Antimetabolites
  Cell cycle sensitive (CCS)—
    primarily in the S phase
Plant alkaloids
  Vinblastine and vincristine—
    CCS—primarily in the
    M phase
  Ectoposide—CCS—S and early
    G2 phase
  Paclitaxel—spindle poison
Antibiotics
  Bleomycin—CCS—primarily in
    G2 phase
  Doxyrubicin, dactinomycin, and
    mitomycin—cell cycle non-
    sensitive
Alkylating agents and hormones—
    cell cycle nonspecific
    (CCNS)

Cardiovascular and Pulmonary Systems

Drugs used in congestive heart
    failure
  Positive inotropes
  Diuretics
  ACE inhibitors
  PDE inhibitors
  Vasodilators

Antianginals
  Calcium channel blockers
  Nitrates
  β-adrenergic blockers
Antiarrhythmics
  Sodium channel blockers
  β-adrenergic blockers
  Potassium channel blockers
  Calcium channel blockers
  Adenosine
  Digoxin
Antihypertensives
  Diuretics
  Adrenergic receptor blockers
  Vasodilators
  Angiotensin antagonists
Antihyperlipidemics
  Resins
  HMG-CoA reductase inhibitors
  Niacin
  Gemfibrozil
Drugs used in clotting disorders
  Clot reducers
    Anticoagulants
    Antiplatelet agents
    Thrombolytics
  Clot facilitators
    Replacement factors
    Plasminogen inhibitors
Antiasthmatics
  Bronchodilators
  Anti-inflammatories
  Leukotriene antagonists

Drug	Adverse Drug Reaction
Digoxin	Arrythmias, visual aberrations
Nitrates	Tachycardia, headaches, and tolerance
Verapamil	Constipation
β-adrenergic blockers	Bradycardia and asthma
Quinidine and sotalol	Torsades-like arrhythmia
Procainamide	Lupus-like reaction
Amiodarone	Pulmonary fibrosis, thyroid dysfunction, and constipation
Prazosin	First-dose orthostatic hypotension
Clonidine	Rebound hypertension on acute drug cessation
Methyldopa	Positive Coombs test
Guanethidine	Orthostatic hypotension
Reserpine	Depression
Hydralazine	Lupus-like syndrome
Minoxidil	Hirsutism, marked salt and water retention
ACE inhibitors	Dry cough, contraindicated in renal disease
Resins	Bloating
HMG-CoA reductase inhibitors	Severe muscle pain
Niacin	Flushing

Central Nervous System

Antipsychotics
  Phenothiazines
  Thioxanthines
  Butyrophenones
  Heterocyclics
  Antimanics

Adverse Drug Reactions of Antipsychotics

Extrapyramidal effects—haloperidol, fluphenazine

Tardive dyskinesia

Atropine-like effects—thioridizine, chlorpromazine, clozapine

Orthostatic hypotension

Hyperprolactinemia

Amenorrhea-galactorrhea syndrome

Neuroleptic malignant syndrome

Agranulocytosis—clozapine

Nephrogenic diabetes insipidus—lithium

Antidepressants

Monoamine oxidase (MAO) inhibitors

Tricyclics

Heterocyclics

Selective serotonin reuptake inhibitors

α_2-adrenergic blockers

Adverse Drug Reactions of Antidepressants

Combination of MAO inhibitors and fluoxetine—serotonin syndrome

MAO inhibitors and foods containing tyramine—hypertensive crisis

Opiates
  Agonists
  Mixed agonists
  Antagonists

Anxiolytics
  Benzodiazepines

Barbiturates

Ethanol

Antiparkinsonians
  Dopamine antagonists
  MAO inhibitors
  Antimuscarinics

Adverse Drug Reactions of Antiparkinsonians

Levodopa, bromocryptine—choreoathetosis

Antiepileptics
  Phenytoin
  Carbamazepine
  Valproic acid
  Gabapentin
  Vigabatrin
  Ethosuximide
  Benzodiazepines

Drug	Use
Valproic acid, phenytoin, carbamazine	Grand mal seizures
Ethosuximide, valproic acid	Absence seizures
Valproic acid	Myoclonus
Diazepam, lorazepam	Status epilepticus

Drug	Adverse Drug Reaction
Valproic acid	Neural tube defects
Phenytoin	Nystagmus, gingival hyperplasia, ataxia, hirsutism
Carbamazepine	Diplopia, ataxia
Gabapentin	Movement disorders, behavioral aberrations in children
Vigabatrin	Agitation, confusion, psychosis

Autonomic Nervous System
Location and function of adrenergic
   and cholinergic receptors
Cholinergic agents
   Direct acting
      Nicotinic
      Muscarinic
   Indirect acting
      Organophosphates
      Carbamates
      Quarternary alcohols
Anticholinergic agents
   Antimuscarinic
   Antinicotinic

Ganglionic blockers
Neuromuscular blockers
Adrenergic agents
   Direct acting
      α-adrenergic agonists
      β-adrenergic agonists
   Indirect acting
      Releasers
      Reuptake inhibitors
Antiadrenergic agents
   α-adrenergic blockers
   β-adrenergic blockers

Drug	Use
Edrophonium, pyridostigmine, neostigmine	Myasthenia gravis
Carbachol, pilocarpine, physostigmine, timolol	Glaucoma
Tacrine, donepezil	Alzheimer's disease
Pralidoxime	Organophosphate poisoning antidote
Scopalamime	Motion sickness
Ipratropium	Chronic obstructive pulmonary disease
Albuterol	Asthma
Dopamine	Cardiogenic shock
Dobutamine	Cardiogenic shock and congestive heart failure
Ephedrine, oxymetazoline, phenylephrine	Nasal congestion
Phentolamine, phenoxybenzamine	Pheochromocytoma
Prazosin	Hypertension
Beta blockers	Angina, hypertension, arrhythmias, and myocardial infarction
Epinephrine	Anaphylaxis
Tropicamide	Mydriasis and cycloplegia

Drug	Adverse Drug Reaction
Muscarinics	Nausea, vomiting, diarrhea, salivation, sweating, cutaneous vasodilation, and bronchial constriction
Nicotinics	Convulsions, respiratory paralysis, and hypertension
Cholinesterase inhibitors	Signs of muscarinic and nicotinic toxicities
Antimuscarinics	Hyperthermia due to blockage of sweating mechanisms, decreased salivation and lacrimation, acute-angle-closure glaucoma in the elderly, urinary retention, constipation, blurred vision, delirium, and hallucinations
Antinicotinics	Respiratory paralysis
Adrenergics	Marked increase in blood pressure, tachycardia
α-adrenergic blockers	Orthostatic hypotension, reflex tachycardia
β-adrenergic blockers	Bradycardia, atrioventricular blockade, negative inotropy, bronchiolar constriction, hypoglycemia

Local Control Substances

Histamine antagonists
  H_1
  H_2
Serotonin agonists
  $5\text{-}H_1$
  Serotonin-selective reuptake inhibitors
Serotonin antagonists
  $5\text{-}HT_2$
  $5\text{-}HT_3$
Ergot alkaloids
  CNS
  Uterus
  Vessels

Eicosonoid agonists
  Prostaglandins
  Prostacyclins
  Thromboxanes
  Leukotrienes
Eicosonoid antagonists
  Corticosteroids
  Nonsteroidal anti-inflammatory drugs (NSAIDs)
  Leukotriene antagonists

Drug	Use
Histamine antagonists	
H_1	Allergies
H_2	Acid-peptic disease
Serotonin agonists	
Sumatriptan	Acute migraine and cluster headaches
Serotonin antagonists	
Ketanserin, cyproheptadine,	Carcinoid tumors
and phenoxybenzamine	
Ondansetron	Postoperative vomiting and vomiting
	associated with cancer chemotherapy
Ergot alkaloids	
Ergotamine	Acute migraine headache
Methysergide and ergonovine	Prophylactic use for migraine headaches
Ergonovine and ergotamine	Reduction of postpartum bleeding
Bromocriptine and pergolide	Reduction of prolactin secretion
Eicosanoid agonists	
Misoprostol	Abortifacient, prevention of ulcers in
	combination with NSAIDs therapy
PGE_1	Maintain patency of ductus arteriosus
Alprostadil	Erectile dysfunction
Corticosteroids	Inhibition of arachidonic acid production
NSAIDs	Closure of patent ductus arteriosus
Indomethacin	Asthma
Leukotriene antagonists	

Drug	Adverse Drug Reaction
Histamine receptor antagonists	
H_1	Drowsiness
H_2-cimetidine	Inhibitor of drug-metabolizing enzymes
Serotonin antagonists	
Ketanserin	α and H_1 antagonism
Ondansetron	Diarrhea and headache
Ergot alkaloids	Ischemia and gangrene, fibroplasia of
	connective tissue, uterine contractions,
	and hallucinations

Renal System

Diuretics effecting salt and water
   excretion
   Osmotic
   Carbonic anhydrase inhibitors
   Loop diuretics
   Thiazides

Potassium-sparing diuretics
Drugs effecting water excretion
   Osmotic
   ADH agonists
   ADH antagonists

Drug	Use
Loop diuretics	Congestive heart failure and pulmonary edema, ascites
Thiazides	Hypertension, congestive heart failure, renal calcium stones
Osmotics	Increasing urine flow, decreasing intracranial pressure
Potassium-sparing diuretics	Diminishing potassium wasting from other diuretics
ADH	Pituitary diabetes insipidus

Drug	Adverse Drug Reaction
Loop diuretics	Hypokalemia, ototoxicity
Thiazides	Hypokalemia, hyperglycemia, hyperuricemia, hyperlipidemia
Potassium-sparing diuretics	Hyperkalemia
Spironolactone	Gynecomastia and antiandrogenic effects
ADH	Hyponatremia

Gastrointestinal System and Nutrition

Gastrointestinal tract ulcers
   Antacids
   Polymers (sucralfate)
   Proton pump inhibitors
   Antibiotics

Gastrointestinal motility
   promotors
Antiemetics
   H$_2$ antagonists
   Phenothiazines
   5-HT inhibitors

Pancreatic replacement enzymes
Laxatives
Irritants
Bulk formers

Stool softeners
Lubricants
Sulfasalazine
Antidiarrheals

Drug	Use
Ondansetron	Antiemetic in cancer chemotherapy
Omeprazole	Zollinger-Ellison syndrome

Endocrine System

Androgens
  Testosterone

Antiandrogens
  GnRH analogs
  Steroid synthesis inhibitors
  5α reductase inhibitors
  Testosterone receptor inhibitors

Estrogens

Progesterones

Corticosteroids
  Glucocorticoids
  Mineralocorticoids

Corticosteroid antagonists

Receptor antagonists

Synthetic inhibitors

Thyroid hormones
  Thyroxine
  Triiodothyronine

Antithyroid hormones
  Thioamides
  Iodide

Radioactive iodide
Ipodate

Antidiabetics
  Insulin
  Oral hypoglycemics
    Sulfonylureas
    Biguanides
    Thiazolidinediones
    Acarbose

Hyperglycemics

Bone mineral metabolism agents
  Parathyroid hormone
  Vitamin D
  Calcitonin—Paget's disease and
  hypercalcemia
  Estrogens
  Glucocorticoids
  Biphosphonates—post-
  menopausal osteoporosis
  Fluoride
  Plicamycin—Paget's disease

Drug	Adverse Drug Reaction
Androgens	Masculinizing effects
Estrogens	Breakthrough bleeding and breast tenderness
Thyroid hormones	Thyrotoxicosis
Glucocorticoids	Adrenal suppression, salt retention, diabetes, osteoporosis
Insulin	Hypoglycemia
Sulfonylureas	Hypoglycemia
Biguanides	Diarrhea, lactic acidosis in renal or hepatic insufficiency and anoxic states
Thiazolidinediones	Possible hepatotoxicity
Etidronate	Esophageal irritation
Fluoride	Ectopic bone formation, exostosis
Vitamin D	Nephrocalcinosis

Toxicology

Air pollutants
  Carbon monoxide
  Sulfur dioxide
  Nitrogen oxides
  Ozone
Solvents
  Halogenated hydrocarbons
  Aromatic hydrocarbons
Insecticides
  Chlorinated hydrocarbons
  Cholinesterase inhibitors
  Botanical insecticides

Herbicides
Environmental pollutants
  Dioxins
  Polychlorinated biphenyls
Heavy metals
  Lead
  Arsenic
  Mercury
  Iron

Toxin	Treatment
Carbon dioxide	Removal from exposure and administer oxygen
Sulfur dioxide	Removal from exposure
Aliphatic hydrocarbons	Removal from exposure
Aromatic hydrocarbons	Removal from exposure
Cholinesterase inhibitors	Atropine, pralidoxime
Paraquat	Gastric lavage and dialysis
Lead	Dimercaprol, penicillamine
Arsenic	Dimercaprol, penicillamine
Mercury	Dimercaprol (elemental), penicillamine, dimercaprol (inorganic salts)
Iron	Deferoxamine

High-Yield Facts
in Physiology

CELLULAR PHYSIOLOGY

- Ions, nutrients, and waste material are transported across cell membranes by diffusion, osmosis, and active transport.

Simple diffusion is described by the Fick equation. Carrier-mediated diffusion, called *facilitated diffusion*, is described by the Michaelis-Menton equation (see figure below).

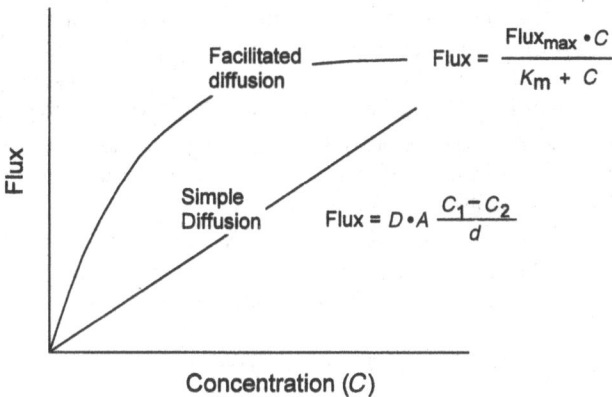

The flow of water through membranes by osmosis is described by the osmotic flow equation. Material dissolved in the water is carried across the membrane by solvent drag.

$$\text{Flow} = \sigma \cdot L \cdot (\pi_1 - \pi_2)$$

σ = reflection coefficient
L = hydraulic conductivity
π = osmotic pressure

The reflection coefficient (σ)

$$\sigma = 1 - \frac{P_{solute}}{P_{water}}$$

is an index of the membrane's permeability to the solute and varies between 0 and 1. Particles that are impermeable to the membrane have a reflection coefficient of 1. Particles that are freely permeable to the membrane have a reflection coefficient of 0. The osmotic pressure (in units of mmHg) is calculated with the van't Hoff equation:

$$\pi = c \cdot R \cdot T$$

Cells shrink when placed in hypertonic solutions and swell when placed in hypotonic solutions. *Tonicity* is the concentration of nonpermeable particles. The following equation is used to calculate the steady state volume of the cell:

$$\pi_{initial} \cdot V_{initial} = \pi_{final} \cdot V_{final}$$

Active transport processes may be primary or secondary. Primary active transport processes, such as the Na-K pump, use the energy derived from the hydrolysis of ATP to transport materials against their electrochemical gradient. Secondary active transport processes, such as those that transport glucose and amino acids into renal or intestinal epithelial cells, use the energy from the Na^+ electrochemical gradient.

- All cells have membrane potentials. The magnitude of the membrane potential is determined by the membrane permeability and ion concentration gradient of the ions that are permeable to the membrane.

In the resting state, the membrane is primarily permeable to K^+, and, therefore, the resting membrane potential is close to the equilibrium potential for K^+. The equilibrium potential is calculated with Nernst's equation:

$$\left(E_{ion} = -60 \cdot \log \frac{Conc_{in}}{Conc_{out}} \right)$$

The resting potential is calculated with the transference equation:

$$\left(\begin{array}{l} E_M = T_K \cdot E_{Na} + T_{Na} \cdot E_{Na} \\ T_K = \frac{g_K}{g_{Na} + g_K}; \ T_{Na} = \frac{g_{Na}}{g_{Na} + g_K} \end{array} \right)$$

or with the Goldman equation:

$$E_M = -61 \cdot \log \frac{P_{Na} \cdot [Na]_{in} + P_K \cdot [K]_{in}}{P_{Na} \cdot [Na]_{out} + P_K \cdot [K]_{out}}$$

Action potentials are produced by voltage and time-dependent gates covering ion selective channels. Nerve and skeletal muscle membranes contain Na^+ and K^+ ion selective channels (see figure on p 3).

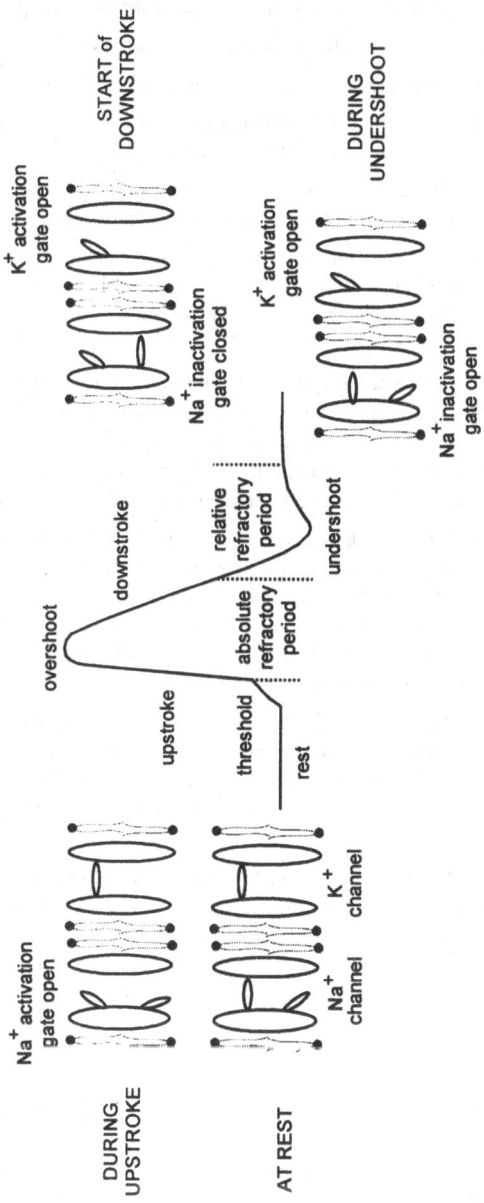

DURING UPSTROKE

Na⁺ activation gate open

K⁺ channel

AT REST

Na⁺ channel

START of DOWNSTROKE

K⁺ activation gate open

Na⁺ inactivation gate closed

DURING UNDERSHOOT

K⁺ activation gate open

Na⁺ inactivation gate open

overshoot

upstroke   downstroke

threshold   absolute refractory period   relative refractory period

rest   undershoot

- Cardiac muscle membranes contain Na^+, K^+, and Ca^{2+} ion selective channels.
- The upstroke (phase 0) is produced by the activation of Na^+ channels.
- The initial repolarization (phase 1) is produced by inactivation of Na^+ channels.
- The plateau (phase 2) is caused by the activation of Ca^{2+} channels and the closing of inward rectifying (anomalous) K^+ channels.
- The downstroke (phase 3) is caused by the activation of the delayed rectifier K^+ channels and the inactivation of Ca^{2+} channels (see figure on p 5).
- Increasing extracellular K^+ concentration depolarizes the membrane and reduces its excitability. Excitability is reduced because the depolarization inactivates Na^+ channels. The reduced excitability can lead to muscle weakness and cardiac arrhythmias.
- Synaptic transmission is used to transmit information from one cell to another cell. The synaptic transmitter, released from the presynaptic cell by exocytosis, diffuses across a synaptic cleft and binds to a receptor on the postsynaptic cell. The effect produced on the postsynaptic cell depends on the identity of the synaptic transmitter and the receptor (see figure on p 6).

   Acetylcholine, which binds to the end plate of skeletal muscle cells, and glutamate and GABA, which bind to the postsynaptic membranes of many central nervous system membranes, open ion selective channels.

   Norepinephrine and acetylcholine, which bind to the postsynaptic membranes of smooth muscle cells, produce their effect by activating a G protein which, in turn, activates an enzyme-mediated response.

- Muscle contraction is produced by repetitive cycling of the myosin cross-bridges on thick filaments. The cross-bridges attach to actin molecules on the thin filaments and cause the thin filaments to slide over the thick filaments toward the center of the sarcomere (skeletal or cardiac muscle) or cell (smooth muscle; see figure on p 7).

   The initiation of contraction is called *excitation-contraction coupling*. In striated muscle, excitation-contraction coupling is initiated when Ca^{2+} binds to troponin. Troponin causes tropomyosin to move, thereby exposing the actin binding site to myosin (see figure on p 8).

   In skeletal muscle, Ca^{2+} is released from the sarcoplasmic reticulum (SR) when the muscle fiber depolarizes.

   In cardiac muscle, Ca^{2+} is released from the SR by the Ca^{2+} that enters the cell during the cardiac action potential.

   In smooth muscle, excitation-contraction coupling is initiated when Ca^{2+} binds to calmodulin.

   The Ca^{2+}-calmodulin complex activates myosin light chain kinase (MLCK) which, in turn, phosphorylates the 20,000-Da myosin light chains (LC_{20}). Cross-bridge cycling begins when the myosin light chains are phosphorylated.

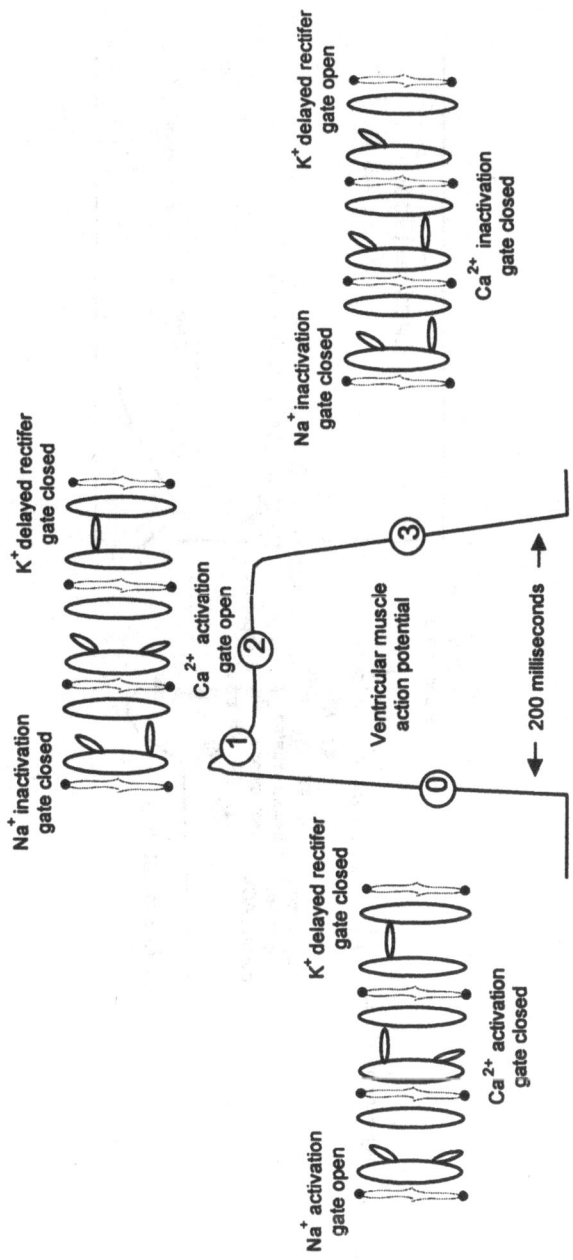

Na+ inactivation
gate closed

K+ delayed rectifier
gate closed

Ca2+ activation
gate open

K+ delayed rectifier
gate open

Na+ inactivation
gate closed

Ca2+ inactivation
gate closed

Na+ activation
gate open

K+ delayed rectifier
gate closed

Ca2+ activation
gate closed

Ventricular muscle
action potential

← 200 milliseconds →

1 2 3 0

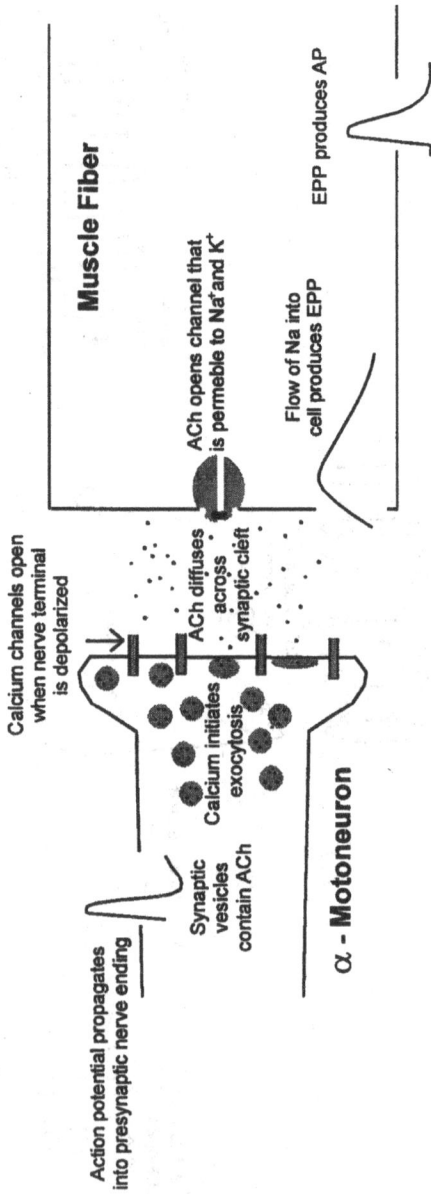

Muscle Fiber

Action potential propagates
into presynaptic nerve ending

Calcium channels open
when nerve terminal
is depolarized

Synaptic
vesicles
contain ACh

Calcium initiates
exocytosis

ACh diffuses
across
synaptic cleft

ACh opens channel that
is permeble to Na⁺ and K⁺

Flow of Na into
cell produces EPP

EPP produces AP

α - Motoneuron

Receptor	Enzyme	Enzyme Effect	Postsynaptic Response
Adrenergic Alpha$_1$ (G$_o$)	Activation of phospholipase C	Formation of IP$_3$ and DAG	IP$_3$ releases calcium from SR DAG activates PKC Calcium activates arteriolar smooth muscle cells
Adrenergic Alpha$_2$ (G$_i$)	Inhibition of adenylyl cyclase	Reduction in cAMP formation	Relaxes GI smooth muscle cells
Adrenergic Beta (G$_s$)	Activation of adenylyl cyclase	Formation of cAMP	cAMP activates PKA which: • Increases Ca^{2+} entry into cardiac muscle cells and increases contractility • Increases sequestration of Ca^{2+} in bronchiole smooth muscle cells and relaxes muscle
Cholinergic Muscarinic (M$_1$)	Activation of phospholipase C	Formation of IP$_3$ and DAG	IP$_3$ releases calcium from SR DAG activates PKC Calcium activates GI smooth muscle cells

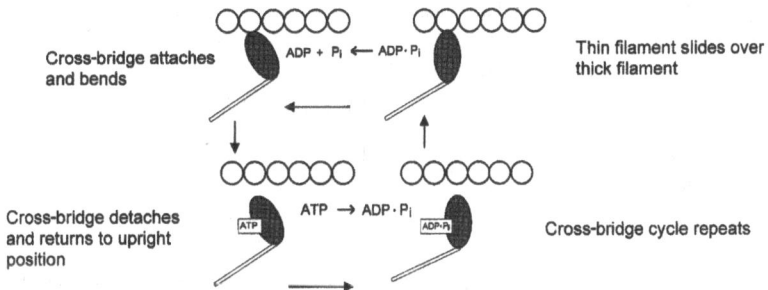

Cross-bridge attaches and bends — ADP + P$_i$ ← ADP·P$_i$ — Thin filament slides over thick filament

Cross-bridge detaches and returns to upright position — ATP → ADP·P$_i$ — Cross-bridge cycle repeats

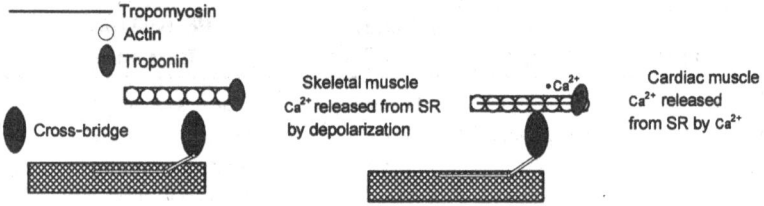

When dephosphorylated, the cross-bridges stay attached (or cycle slowly). The attached, slowly cycling cross-bridges are called latch bridges. Latch bridges allow smooth muscle to maintain force while minimizing energy expenditure.

CARDIAC AND VASCULAR PHYSIOLOGY

- The heart pumps the blood through the circulation. The cardiac output from the heart must be sufficient to perfuse all the organs and maintain a pressure adequate to perfuse the brain (see figure below).

Mean Blood Pressure = Cardiac Output × Total Peripheral Resistance

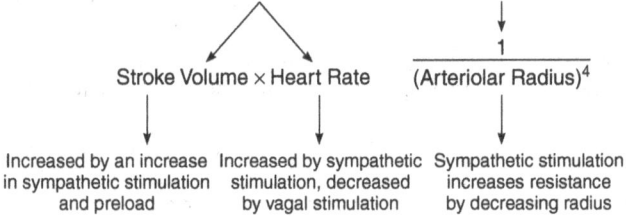

$$\frac{1}{(\text{Arteriolar Radius})^4}$$

Stroke Volume × Heart Rate

| Increased by an increase in sympathetic stimulation and preload | Increased by sympathetic stimulation, decreased by vagal stimulation | Sympathetic stimulation increases resistance by decreasing radius |

Cardiac output can be measured using the Fick equation:

$$CO = \frac{\dot{V}O_2 \text{ (oxygen consumption)}}{(\text{Arterial } O_2 \text{ content} - \text{Venous } O_2 \text{ content})}$$

Resistance can be calculated using the Poiseuille equation:

$$R = \frac{8 \cdot \eta \cdot L}{\pi \cdot r^4}$$

η = viscosity
L = length
r = radius

Because resistance is inversely proportional to the 4th power of the radius, TPR is controlled by small variations in the diameter of the arterioles.

Blood pressure is maintained by the baroreceptor reflex, which responds to a decrease in blood pressure by increasing heart rate, contractility, and TPR, and by decreasing venous compliance.

Blood pressure decreases as blood flows through the circulation. The magnitude of the decrease is proportional to the resistance of each segment of the circulation. The greatest decrease occurs as blood flows through the arterioles. The segments of the circulation are in series with each other (see figure below).

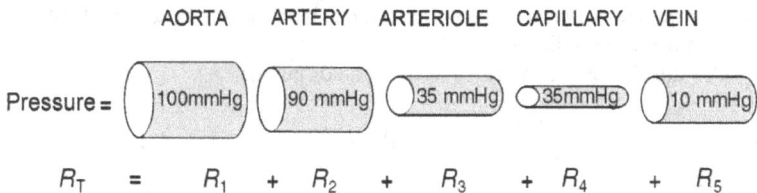

AORTA	ARTERY	ARTERIOLE	CAPILLARY	VEIN

$$\text{Pressure} = \boxed{100\text{mmHg}} \quad \boxed{90\text{ mmHg}} \quad 35\text{ mmHg} \quad 35\text{mmHg} \quad \boxed{10\text{ mmHg}}$$

$$R_T \;=\; R_1 \;+\; R_2 \;+\; R_3 \;+\; R_4 \;+\; R_5$$

The quantity of blood flowing into each organ is inversely proportional to the relative resistance of each organ. For example, at rest approximately 20% of the cardiac output flows through the skeletal muscles. During exercise, when the resistance of the skeletal muscle vessels decreases, over 80% of the cardiac output can flow through the skeletal muscles. The organs in the body are in parallel with each other (see figure below).

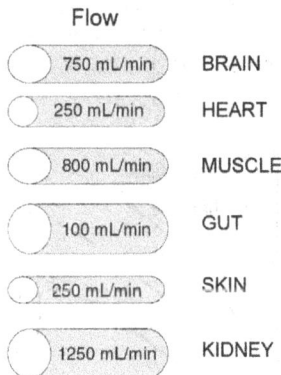

Flow

750 mL/min	BRAIN
250 mL/min	HEART
800 mL/min	MUSCLE
100 mL/min	GUT
250 mL/min	SKIN
1250 mL/min	KIDNEY

$$\frac{1}{R_T} = \frac{1}{R_1} + \frac{1}{R_2} + \frac{1}{R_3} + \frac{1}{R_4} + \frac{1}{R_5} + \frac{1}{R_6}$$

The velocity of blood through a vessel is proportional to the flow of blood through the vessel and inversely proportional to the area of the vessel:

$$v = \frac{Q}{A}$$

Increasing the velocity of blood can change flow from laminar (the fluid moves in a steady, orderly stream) to turbulent (the fluid movement is disorderly).

Turbulent flow produces a sound called a murmur if it occurs in the heart, or a bruit if it occurs in a blood vessel. Flow through stenotic or incompetent heart valves produces cardiac murmurs. Occlusion of blood vessels by a sclerotic plaque, for example, will produce bruits.

Turbulence can be predicted from Reynolds number:

$$N_R = \frac{\rho \cdot D \cdot V}{\eta}$$

ρ = density
D = diameter
V = velocity
η = viscosity

If Reynolds number exceeds 2000–3000, flow is likely to be turbulent.

- Stroke volume (and cardiac output) is dependent on preload, afterload, and contractility.

The relationship between preload and stroke volume (or cardiac output) is represented by a Starling curve. The Starling curve is shifted up and to the left by an increase in contractility or a decrease in afterload. The Starling curve is shifted down and to the right by a decrease in contractility or an increase in afterload (see figure below).

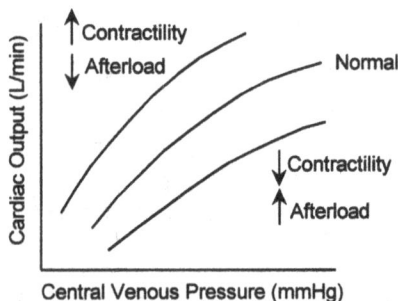

Central Venous Pressure (mmHg)

Preload is dependent on blood volume, venous compliance, and TPR. The relationship between these variables is represented by a vascular function curve. Changes in vascular volume or venous compliance cause a parallel shift in the vascular function curves. Changes in TPR cause the slope of the vascular function curve to change (see figure below).

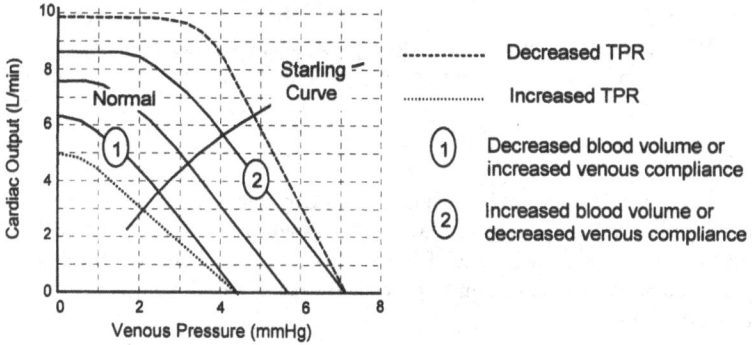

The change in pressure and volume within the heart during one cardiac cycle can be represented by a pressure-volume loop (see figure below).

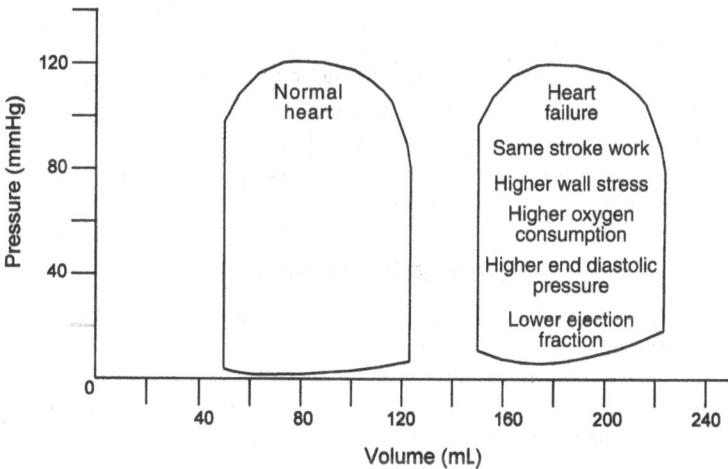

The work required to eject the blood is called the *stroke work*. The stroke work is the product of mean left ventricular systolic pressure and stroke volume:

$$\text{Stroke work} = \text{MLVSP} \cdot \text{SV}$$

Stroke work is equal to the area within the pressure-volume loop. The energy required to eject the blood is dependent on the stroke work and the wall stress.

Wall stress is proportional to systolic pressure and the radius of the ventricle and inversely proportional to the thickness of the ventricular wall (law of Laplace):

$$\text{Wall stress} = \frac{P \cdot r}{\text{Thickness}}$$

Wall stress increases in heart failure because the preload increases to compensate for the decrease in contractility. The increased radius of the enlarged heart causes wall stress to increase, and, therefore, more energy is required to eject blood. If the coronary circulation cannot provide the necessary oxygen, ischemic pain (angina) results.

- Fluid exchange across capillaries is dependent on Starling forces:

$$\text{Filtration} = k_f \cdot [(P_{cap} + \pi_i) - (\pi_{cap} + P_i)]_i$$

Normally, the fluid filtered from the capillaries is returned to the circulation by the lymphatic system. Fluid accumulation in the interstitial space is referred to as *edema*.

Edema can result from:

An increase in capillary permeability, as during an inflammatory response

A decrease in plasma proteins, as during malnutrition

An increase in capillary pressure, as in heart failure (the increased end-diastolic pressure that occurs in heart failure causes the increase in capillary pressure)

A blockage of the lymphatic circulation

PULMONARY PHYSIOLOGY

- Air is moved in and out of the lungs by the movement of the diaphragm and chest. During inhalation, the diaphragm descends and the rib cage moves up and out. The expansion of the lungs creates a negative intra-alveolar pressure (with respect to atmospheric pressure) which draws air into the alveoli. The air moving into the lung with each breath is called the *tidal volume*. The amount of air moving into the lung per minute is called the *minute ventilation*.

Minute ventilation (V_{min}) = Tidal volume (V_T) · Breathing frequency (f)

At the end of inspiration, the air within the conducting airways is called the *anatomical dead space air* and does not contribute to gas exchange. The fresh air entering the alveoli each minute is called the *alveolar ventilation*.

$$\text{Alveolar ventilation } (\dot{V}_A) = (V_T - V_D) \cdot f$$

The physiological dead space is the sum of the anatomical and alveolar dead spaces. It is calculated with Bohr's equation:

$$V_D = V_T \cdot \left(\frac{Faco_2 - Feco_2}{Faco_2} \right)$$

The anatomical dead space (in mL) is approximately equal to the weight (in pounds).

The alveolar dead space (areas of the lung that are ventilated but not appropriately perfused) is typically zero.

- The gas moving in and out of the lungs is measured with spirometry. Tidal volume is the gas moving in and out of the lungs with each breath. The maximum amount of gas that can be expelled from the lungs after breathing in as far as possible is called the *vital capacity*. The maximum amount of gas that can be expelled from the lungs after a normal breath is called the *expiratory reserve volume*. The maximum amount of gas that can be inhaled at the end of a normal breath is called the *inspiratory capacity*. The maximum amount of gas that can be inhaled at the end of a normal inspiration is called the *inspiratory reserve volume* (see figure below).

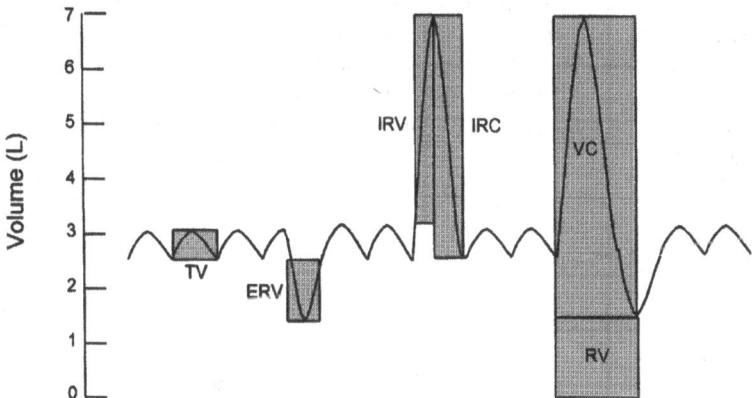

The gas remaining in the lung at the end of maximum expiration (the residual volume) cannot be measured by spirometry. It is measured using the helium dilution technique (see figure below).

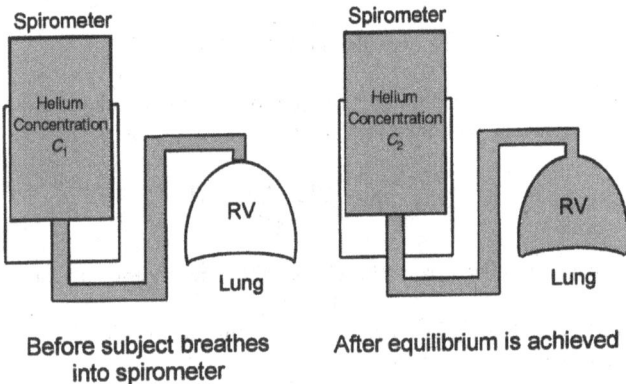

Before subject breathes into spirometer   **After equilibrium is achieved**

A known amount of helium is placed in a spirometer. The subject is connected to the spirometer when the volume in his or her lung is at the residual volume. After equilibration is achieved, the concentration of helium in the spirometer is measured. The following equation is then used to calculate the RV.

$$RV = V_{\text{Spirometer before equilibration}}\left(\frac{C_{\text{Spirometer before equilibration}}}{C_{\text{Spirometer after equilibration}}} - 1\right)$$

The equation calculates whatever volume is in the lung when the subject begins to breathe into the spirometer. Therefore, the helium dilution technique can be used to measure any lung volume. For example, if the subject is connected to the spirometer at the end of a normal expiration, the FRC is calculated.

The *FRC* is the amount of gas in the lung at the end of a normal breath. It equals the RV + ERV.

The *total lung capacity* (TLC) is the amount of gas in the lung after a maximum inspiration. It equals RV + VC.

- The partial pressure of a gas is measured after water has been removed. Therefore, the partial pressure is calculated as a fraction of the atmospheric pressure remaining when the partial pressure of water vapor is subtracted.

$$F_{\text{gas}} \cdot (P_{\text{atm}} - P_{\text{H}_2\text{O}}) = F_{\text{gas}} \cdot (P_{\text{atm}} - 47 \text{ mmHg})$$

The average partial pressure of carbon dioxide in the alveoli is proportional to carbon dioxide production and inversely proportional to alveolar ventilation:

$$Fa_{CO_2} = (\dot{V}_{CO_2}/\dot{V}_a)$$

$$Pa_{CO_2} \propto (\dot{V}_{CO_2}/\dot{V}_a)$$

The average partial pressure of oxygen in the alveoli is calculated using the alveolar gas equation:

$$Pa_{O_2} = Pi_{O_2} - (Pa_{CO_2}/R)$$

$$R = (\dot{V}_{CO_2}/\dot{V}_{O_2})$$

R is the respiratory gas exchange ratio. Under normal circumstances its value depends on metabolism and is equal to 0.8. When the Pi_{O_2} is 100%, the value of R used in the alveolar gas equation is 1.0.

- The lung is an elastic organ that resists stretching. To expand the lungs, the inspiratory muscles must overcome the recoil force of the lungs and the resistance of the airways to airflow. The intrapleural pressure is a measure of the elastic and resistive work done by the inspiratory muscles. Expiration is usually passive. That is, the inspiratory muscles relax and the recoil force of the lungs expels the gas. Gas flow during expiration can be increased by contracting the expiratory muscles. However, the maximum expiratory flow is limited by airway compression (see figure below).

  The work done by the lung can be represented by a pressure-volume loop. In restrictive airway disease (e.g., fibrosis), elastic work increases; in obstructive airway disease (e.g., asthma), resistive work increases (see figure on p 16).

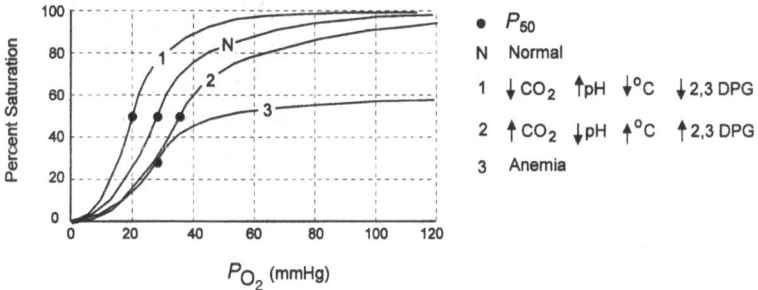

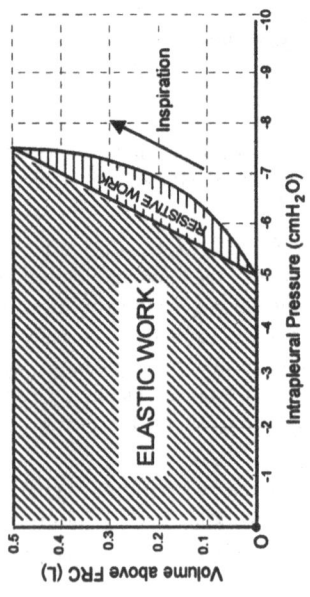

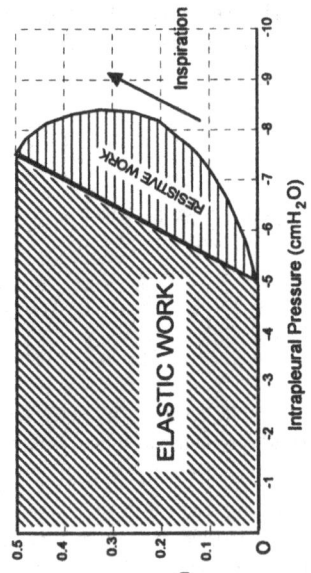

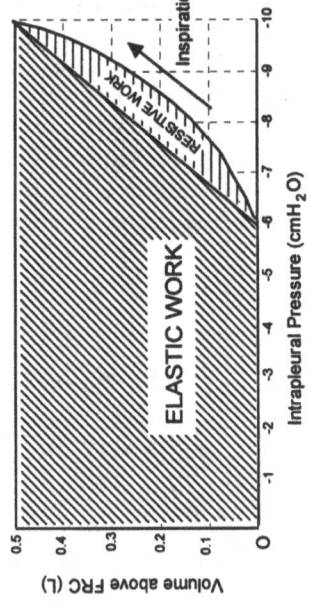

During active expiration, the alveolar pressure is the sum of the recoil force of the lung and the positive intrapleural pressure produced by the expiratory muscles. The intra-airway pressure decreases from the alveoli to the mouth. When the intrapleural pressure equals the intra-airway pressure (the equal pressure point) at a point along the airway where airway collapse can occur, increases in expiratory effort no longer increase expiratory flow (see figure on p 18).

Increases in lung compliance and/or increases in expiratory effort move the equal pressure point closer to the lung.

Maximum expiratory flow rates are lower in both obstructive and restrictive lung diseases.

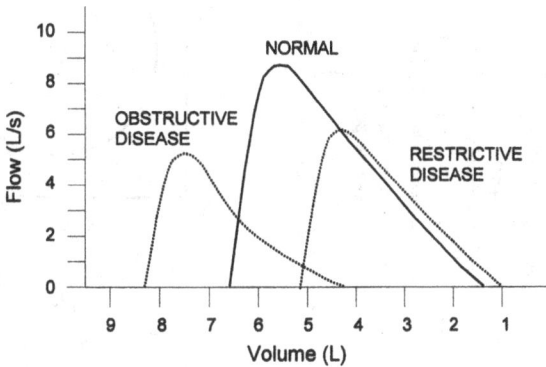

- Gas exchange across the alveolar-capillary membrane is very efficient. Under all but the most serious diffusion abnormalities (such as pulmonary edema), pulmonary capillary blood-gas tensions are in equilibrium with alveolar-gas tensions. Mixed venous blood returning from the systemic circulation has a Pv_{O_2} of 40 mmHg and a Pv_{CO_2} of 46 mmHg. Arterial blood has a Pa_{O_2} of 95 mmHg and a Pa_{CO_2} of 40 mmHg. The central chemoreceptors maintain Pa_{CO_2} at 40 mmHg. The peripheral chemoreceptors will increase ventilation (and reduce Pa_{CO_2} below 40 mmHg) if Pa_{O_2} falls below 60 mmHg or if arterial pH is reduced by a metabolic acidosis.

Most of the oxygen delivered to the blood is bound to hemoglobin (Hb O_2); a much smaller amount is dissolved in the blood. The amount of dissolved oxygen can be calculated if the Pa_{O_2} is known:

$$\text{Dissolved } O_2 \text{ (mL } O_2/100 \text{ mL)} = 0.003 \cdot Pa_{O_2}$$

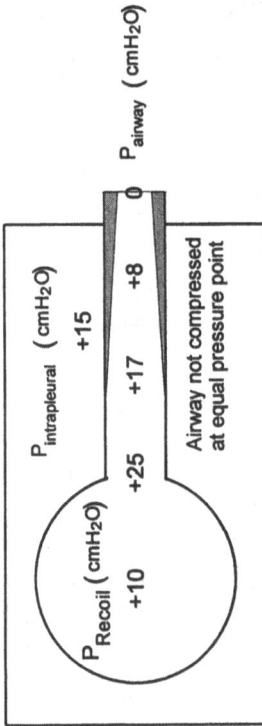

NORMAL

$P_{intrapleural}$ (cmH_2O)

+15

+17

+8

P_{airway} (cmH_2O)

0

P_{Recoil} (cmH_2O)

+10 +25

Airway not compressed
at equal pressure point

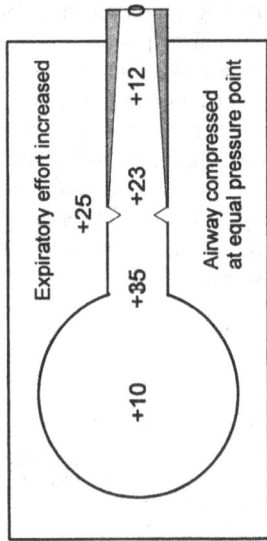

ASTHMA

Expiratory effort increased

+25

+35 +23 +12

0

+10

Airway compressed
at equal pressure point

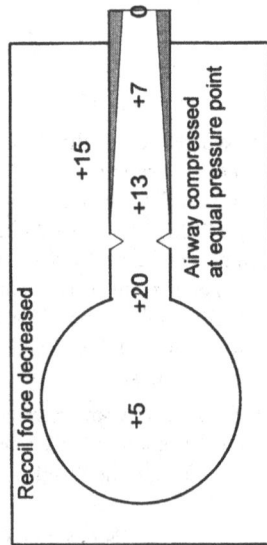

EMPHYSEMA

Recoil force decreased

+15

+20 +13 +7

0

+5

Airway compressed
at equal pressure point

The amount of O_2 combined with hemoglobin can be determined from the oxy-hemoglobin saturation curve or, if the saturation of hemoglobin is known, can be calculated:

$$HbO_2 \text{ (mL } O_2/100 \text{ mL)} = \% \text{ sat} \cdot \text{g of Hb} \cdot 1.34$$

Hemoglobin is 97% saturated at the normal Pa_{O_2} (95 mmHg); 75% saturated at the normal $P\bar{v}_{O_2}$ (40 mmHg); and 50% saturated at 27 mmHg. The Pa_{O_2} at which hemoglobin is 50% saturated is called the P_{50}.

The hemoglobin saturation curve shifts to the right when Pco_2, temperature, H^+ concentration, or 2,3-DPG concentration increases. Shifting the curve to the right makes it easier for O_2 to dissociate from hemoglobin in the tissues. Interestingly, CO causes the Hb saturation curve to shift to the left, making it more difficult for oxygen to dissociate in the tissues.

The $\dot{V}/\dot{Q}$ ratio affects the degree to which hemoglobin is saturated with oxygen. The normal $\dot{V}_A$ of 4 L/min and cardiac output of 5 L/min produces a normal $\dot{V}/\dot{Q}$ ratio of 0.8. Areas of the lung with increased ratios are referred to as *alveolar dead space*; areas of the lung with decreased ratios are referred to as *intrapulmonary shunts*. Because areas of the lung with low $\dot{V}/\dot{Q}$ ratios cannot compensate for areas of the lung with high $\dot{V}/\dot{Q}$ ratios, the $\dot{V}/\dot{Q}$ ratio abnormalities result in low arterial Po_2 and lead to a large A-a gradient for oxygen (see figure below).

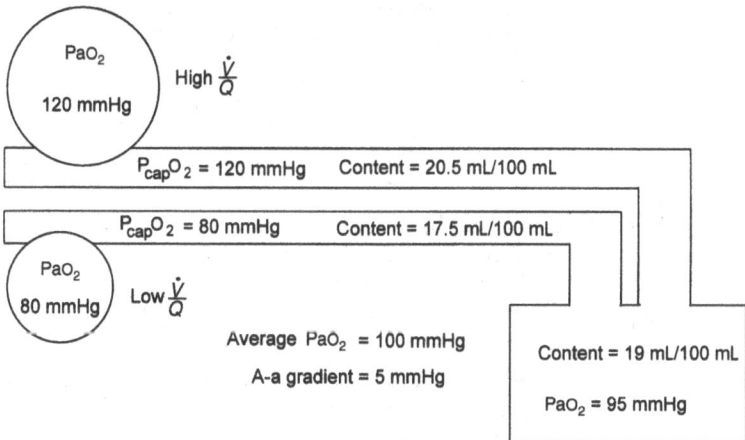

GASTROINTESTINAL PHYSIOLOGY

- The first stage in the digestion and absorption of food is chewing and swallowing. Chewing and swallowing can be initiated voluntarily or involuntarily. Chewing breaks food into small pieces and mixes them with salivary secretions. Swallowing is coordinated by a swallowing center in the brainstem. During the oral phase of swallowing, the tongue pushes the food into the pharynx; during the pharyngeal phase, peristaltic contractions and relaxation of the upper esophageal sphincter (UES) allow the food to enter the esophagus; during the esophageal phase, the lower esophageal sphincter (LES) relaxes and the food is propelled into the stomach by primary peristalsis. A secondary peristaltic wave, initiated by the presence of food in the smooth muscle, clears the esophagus of any food not propelled into the stomach by primary peristalsis.

- The stomach breaks food into small pieces and mixes the pieces with gastric secretions to produce a paste-like material called *chyme*. Liquids and chyme are forced through the pylorus by a rise in gastric pressure. Liquids empty from the stomach in one half-hour. Solids cannot pass through the pyloric sphincter until they are broken into small pieces (less than 1 mm^3), and, therefore, emptying of solids takes from 1 to 3 h.

Gastric Functions		
Function	**Orad Stomach**	**Caudad Stomach**
Motility	Tone	
	Receptive relaxation	Trituration
	Accommodation	Emptying of solids
	Emptying of liquids	Reflux barrier
Secretion	HCl	
	Intrinsic factor	
	Mucus	Gastrin
	Pepsinogen	

Gastric emptying is slowed by the enterogastric reflex and the release of an inhibitory hormone called enterogastrone. The reflex and the secretion of enterogastrone are evoked by the presence of acid or fats in the duodenum.

The secretion of gastric acid by the parietal cells is regulated by paracrine (histamine), neural (vagus nerve), and hormonal (gastrin) influences.

- The intestine is responsible for the digestion and absorption of food and nutrients. During the digestive phase, food is slowly moved along the intestine by segmentation. During the interdigestive phase, the intestine is cleared of any nonabsorbed particles by the migrating motor complex.

Intestinal Phase of Digestion

Stimulus	Mediator	Response
Partially digested nutrients	CCK	Pancreatic enzyme secretion
		Gallbladder emptying
		Decreased gastric emptying
	Enterogastrone and enterogastric reflex	Decreased gastric emptying
		Decreased gastric acid secretion
Acid	Secretin	Pancreatic/hepatic HCO_3^- secretion
		Decreased gastric acid secretion
Osmoles or distension	Enterogastrone and enterogastric reflex	Decreased gastric emptying
		Decreased gastric acid secretion

Sodium Transport Mechanisms in the Small Intestine

Nutrient-Coupled	Na^+/H^+ Exchange	Neutral	Electrogenic
Small intestine	Jejunum	Ileum and colon	Colon

Functions of Small Intestine

	Duodenum	Jejunum	Ileum
Motility		Segmentation (digestive period)	
		Migrating motor complex (interdigestive period)	
Secretion	CCK and secretin		HCO_3^-
Digestion		Intraluminal and brush border	
		Nutrients, water, and electrolytes	
Absorption	Iron and folate	Folate	Bile acids and vitamin B_{12}

Nutrient Digestion and Absorption		
Nutrient	**Digestion**	**Absorption**
Carbohydrates	Intraluminal (salivary, pancreatic amylase)	Active (glucose and galactose)
	Intestinal brush border enzymes	Passive (fructose)
Proteins	Intraluminal (gastric pepsin, pancreatic peptidases)	
	Intestinal brush border enzymes	Active (amino acids and di- and tripeptides)
	Intracellular (cytosolic peptidases)	
Lipids	Intraluminal (gastric and pancreatic lipase)	
	Bile salts required for micelle formation and solubilization of lipids	Passive Intracellular resynthesis Chylomicron formation
	Colipase essential for lipase hydrolysis of micellar lipids	

Fats, proteins, and carbohydrates are digested by intestinal enzymes. Bile is necessary for the digestion and absorption of fats. However, the amount of bile acids emptied into the proximal small intestine from the gallbladder is insufficient for complete fat digestion and absorption. Receptor-mediated active transport of bile acids in the terminal ileum returns the bile acids via the portal blood to the liver for secretion into the small intestine (this circulation of bile is called the *enterohepatic circulation*). Approximately 95% of the bile acid pool is recirculated from the intestine and about 5% is lost in the stool.

Water absorption is caused by osmotic forces generated by active sodium absorption. The source of water is both exogenous (oral input) and endogenous (GI tract secretion) and averages 8 to 10 L/day. Generally, less than 0.2 L/day is eliminated in the stool. The majority of water absorption occurs in the duodenum and jejunum. The colon exhibits a limited capacity to absorb water (4–6 L/day).

RENAL AND ACID-BASE PHYSIOLOGY

- The kidney is responsible for maintaining the constancy and volume of the extracellular fluid. The functional unit of the kidney is the glomerulus and its associated nephron. Each day, 160 to 180 L of fluid are filtered into the approximately one million nephrons in the human kidney. The glomerular filtration rate (GFR) is dependent on the Starling forces:

$$GFR = k_f \cdot [(P_{cap} + \pi_{BC}) - (\pi_{cap} + P_{BC})]$$

The amount of material filtered into the proximal tubule is called the *filtered load.* Approximately 20% of the plasma flowing through the glomerulus (RPF) is filtered into the proximal tubule:

$$\text{Filtration fraction} = \frac{GFR}{RPF}$$

The relative quantity of material excreted by the kidney (the renal clearance) is expressed as the volume of plasma that is completely cleared of the material by the kidney:

$$\text{Renal clearance} = (U_{conc} \cdot \dot{V})/P_{conc}$$

U_{conc} = Urinary concentration of material

P_{conc} = Plasma concentration of material

$\dot{V}$ = Urinary flow rate

If a material is filtered but not reabsorbed or secreted, its renal clearance will be equal to the GFR. The clearances of creatinine or inulin are used clinically to measure GFR.

If a material is completely cleared from the plasma during its passage through the kidney by a combination of filtration and secretion, its renal clearance will be equal to the RPF. The clearance of PAH is used clinically to measure RPF:

$$\text{Renal blood flow (RBF)} = \frac{RPF}{(1 - \text{hematocrit})}$$

- The proximal tubule is responsible for reabsorbing most of the material filtered from the glomerulus.

Material	% Reabsorbed	Mechanism
Na^+	60–70	Na/H exchange
		Na-nutrient cotransport
		Diffusion
K^+, urea, Cl^-	60–70	Diffusion and solvent drag
Glucose, amino acids	100	Na-nutrient cotransport
Phosphate	90	Na-nutrient cotransport
Bicarbonate	85	Indirectly via Na/H exchange

- The loop of Henle is responsible for producing a dilute filtrate. It reabsorbs approximately 25% of the salt and 15% of the water filtered from the glomerulus. The filtrate flowing from the loop of Henle to the distal convoluted tubule has a Na^+ concentration of approximately 100 meq/L.

- The distal nephron is responsible for regulating salt and water balance. Na^+ balance is regulated by aldosterone and atrial natriuretic peptide (ANP). Water balance is regulated by antidiuretic hormone (ADH), which is also called arginine vasopressin (AVP). K^+ balance is regulated by aldosterone.

    Aldosterone increases Na^+ reabsorption and K^+ secretion by the principle cells of the cortical and medullary collecting ducts. Aldosterone acts on the cell nucleus, increasing Na^+ conductance of the apical membrane (which, by allowing more Na^+ to enter the cell, increases Na^+ reabsorption), the number of Na^+-K^+-ATPase pump sites (which, by increasing intracellular K^+ concentration, increases K^+ secretion), and the concentration of mitochondrial enzymes.

    ANP decreases Na^+ reabsorption by the renal epithelial cells of the medullary collecting ducts.

    ADH increases water reabsorption by the principle cells of the cortical and medullary collecting ducts. ADH upregulates the number of water channels on the apical membrane of the epithelial cells by a cyclic-AMP-dependent process.

- The extracellular osmolarity is controlled by ADH.

    Increases in osmolarity stimulate the release of ADH from the posterior pituitary gland. ADH returns osmolarity toward normal by decreasing the amount of water excreted by the kidney. When osmolarity is decreased, ADH release is decreased and osmolarity is returned toward normal by increased water excretion.

    ADH also is secreted in response to low blood pressure. Under these conditions, reabsorption of water by the kidneys can make the extracellular fluid hypotonic.

- The extracellular volume is controlled by the salt content of the extracellular fluid. Salt content is controlled by aldosterone and ANP. Extracellular volume is monitored by low-pressure baroreceptors within the thoracic venous vessels and the atria and by pressure receptors within the afferent arteriole.

    Aldosterone secretion is controlled by the renin-angiotensin system. Renin is released from the juxtaglomerular cells (JG cells) in response to (1) decreased perfusion pressure within the afferent arteriole, (2) sympathetic stimulation of the JG cells, and (3) decreased Cl^- concentration in fluid bathing the macula densa. Renin catalyzes the conversion of angiotensinogen to angiotensin I. Angiotensin I is converted to angiotensin II (AII) by angiotensin-converting enzyme (ACE) located within the lung. AII stimulates aldosterone secretion from the adrenal cortex gland.

    ANP release is controlled directly by stretch receptors within the right atrium.

- Extracellular K^+ is controlled by aldosterone. Increases in extracellular K^+ stimulate the secretion of aldosterone, causing K^+ secretion to increase. K^+ transport into cells is increased by epinephrine and insulin.
- Each day, approximately 15,000 mmol of volatile acid (CO_2) and 50 to 100 meq of fixed acid (hydrochloric acid, lactic acid, phosphoric acid, sulfuric acid, etc.) are produced by metabolism. The pH of the extracellular fluid is maintained by buffering the acid as it is formed and excreting the acid over time. CO_2 is rapidly excreted by the lungs. The kidneys require approximately 24 h to excrete the fixed acids.

$$CO_2 + H_2O \xrightarrow{CA} H_2CO_3 \longleftrightarrow H^+ + HCO_3^-$$

When CO_2 is added to water, it forms H^+ and HCO_3^-.
The pH of plasma is calculated with the Henderson-Hasselbalch equation:

$$pH = 6.1 + \log \frac{HCO_3^-}{0.03 \cdot P_{CO_2}}$$

The H^+ concentration of the plasma is calculated with the Henderson equation:

$$H^+ = 24 \cdot \frac{P_{CO_2}}{HCO_3^-}$$

Plasma P_{CO_2} is normally maintained at 40 mmHg by the central chemoreceptors. Increases in P_{CO_2} cause an increase in ventilation; decreases in P_{CO_2} cause a decrease in ventilation. The respiratory system rapidly eliminates all of the CO_2 produced by metabolism.

Plasma HCO_3^- concentration is normally maintained at 24 meq/L by the kidneys. HCO_3^- is an important buffer for fixed acids produced by metabolism.

$$H^+ + HCO_3^- \longleftrightarrow H_2CO_3 \xrightarrow{CA} CO_2 + H_2O$$

The HCO_3^- lost during the buffering process is replaced by the distal nephron as the fixed acid (H^+ + anion$^-$) is excreted. The replaced HCO_3^- is called *new bicarbonate*.

The H^+ secreted into the urine is buffered by titratable acids (mostly phosphate) and ammonia. The following equation is used to calculate the net acid excretion:

$$\text{Net acid excretion} = ([\text{Titratable acids}] + [NH_4^+] - [HCO_3^-]) \cdot \dot{V}$$

- Acid-base disorders result from failure of the respiratory system and kidneys to maintain P_{CO_2} and HCO_3^- at their normal levels.

Disturbance	Examples	Compensation	Blood-Gas Profile	
Respiratory acidosis	Depression of the respiratory centers	Kidney produces new HCO_3^-	Pa_{CO_2}	↑ (cause)
			pH	↓ (result)
			HCO_3^-	↑ (compensation)
	Respiratory muscle fatigue			
Respiratory alkalosis	Hypoxemia, anxiety	Kidney excretes HCO_3^-	Pa_{CO_2}	↓ (cause)
			pH	↑ (result)
			HCO_3^-	↑ (compensation)
Metabolic acidosis	Excessive fixed-acid production, as in diabetes	Hyperventilation	HCO_3^-	↓ (cause)
			pH	↓ (result)
			Pa_{CO_2}	↓ (compensation)
	Failure of kidney to excrete H^+, as in renal tubular acidosis			
Metabolic alkalosis	Loss of acid, as in excessive vomiting	Hypoventilation	HCO_3^-	↑ (cause)
			pH	↑ (result)
			Pa_{CO_2}	↑ (compensation)
	Movement of H^+ into cells, as in hypokalemia			

NEUROPHYSIOLOGY

- Sensory receptors (touch, pain, temperature, smell, taste, sound, and sight) are activated by environmental stimuli. The stimulus produces a receptor potential; the magnitude of the receptor potential is proportional to the stimulus. The receptor potential produces a train of action potentials. Tonic receptors fire as long as the stimulus is present and encode intensity. Phasic receptors slow down or stop firing during the presentation of the stimulus and encode velocity.

- Sounds are detected by the hair cells within the organ of Corti of the inner ear. The organ of Corti consists of the hair cells and an overlying membrane called the tectorial membrane to which the cilia of the hair cells are attached. Sounds entering the outer ear cause the tympanic membrane to vibrate. Vibration of the tympanic membrane causes the middle ear bones (malleus, incus, and stapes) to vibrate, which in turn causes the fluid within the inner ear to vibrate.

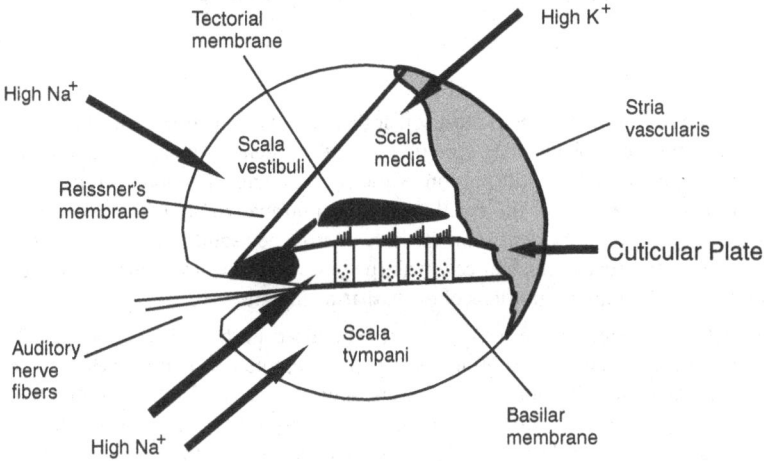

The inner ear is divided into three chambers (scala vestibuli, scala media, and scala tympani). The scala vestibuli is separated from the scala media by Reissner's membrane; the scala media and the scala tympani are separated by the basilar membrane. The stapes is attached to the membrane of the oval window, which separates the middle ear from the scala vestibuli. The scala tympani is separated from the middle ear by the round window. The organ of Corti sits on the basilar membrane. The fluid within the scala vestibuli and scala tympani (perilymph) is similar to interstitial fluid; the fluid within the scala media (endolymph) resembles intracellular fluid in that it contains a high concentration of K^+.

Vibration of the stapes causes the fluid within the scala tympani to vibrate, which in turn causes the basilar membrane to vibrate. Vibration of the basilar membrane causes the cilia to bend back and forth. Bending the stereocilia toward the kinocilium causes K^+ channels on the hair cells to open; bending the stereocilia away from kinocilium causes K^+ channels to close. Auditory hair cells are unusual because they are depolarized by the flow of K^+ into the cell. K^+ can flow into the hair cells because the endolymph surrounding the apical portions of the hair cells contains a high K^+ concentration.

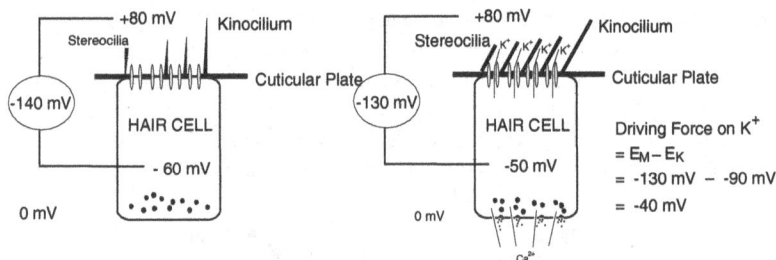

The basilar membrane is most stiff at the base of the cochlea (near the middle ear) and most compliant at the apex of the cochlea. High-frequency sounds cause a greater vibration of the stiff portion of the cochlea, and, therefore, the hair cells located near the base of the cochlea transmit information about high-frequency sounds to the auditory cortex. Similarly, low-frequency sounds are transmitted to the auditory cortex by the hair cells near the base of the cochlea, which are located on the more compliant portions of the basilar membrane.

• Light is detected by the rods and cones contained in the retina of the eye. The retina contains five types of neurons: photoreceptors (rods and cones), bipolar cells, ganglion cells, horizontal cells, and amacrine cells. Light rays from distant objects are normally focused on the photoreceptors by the cornea and the relaxed lens. When objects are brought closer to the eye, they are kept focused on the retina by the accommodation reflex, which causes the refractive power of the lens to increase. The rods and cones contain a visual pigment, called rhodopsin, which absorbs light energy. Rhodopsin contains two components: opsin, which determines the wavelength of light absorbed by rhodopsin, and retinal, which undergoes isomerization by light.

The photoreceptors are unusual because they hyperpolarize when they are stimulated by light. When the rods and cones are not stimulated, they are depolarized by the flow of Na^+ into the cell through Na^+ channels held in the open state by cGMP. The photoisomerization of retinal from its 11-cis form to its all-trans form activates rhodopsin, which in turn activates a G protein called transducin. Activated transducin activates a cGMP esterase. Hydrolysis of cGMP causes Na^+ channels on the rod and cone outer segments to close, which produces the membrane hyperpolarization.

The neurotransmitter keeps the bipolar cells and, therefore, the ganglion cells in a polarized and relatively quiescent state. Hyperpolarization of the photoreceptors stops the release of an inhibitory neurotransmitter, which in turn causes bipolar cells to depolarize. The bipolar cells stimulate ganglion cells, which in turn convey information about the light stimulus to the visual cortex. The ganglion cells are the only cells in the retina to produce an action potential. Their axons form the optic nerve.

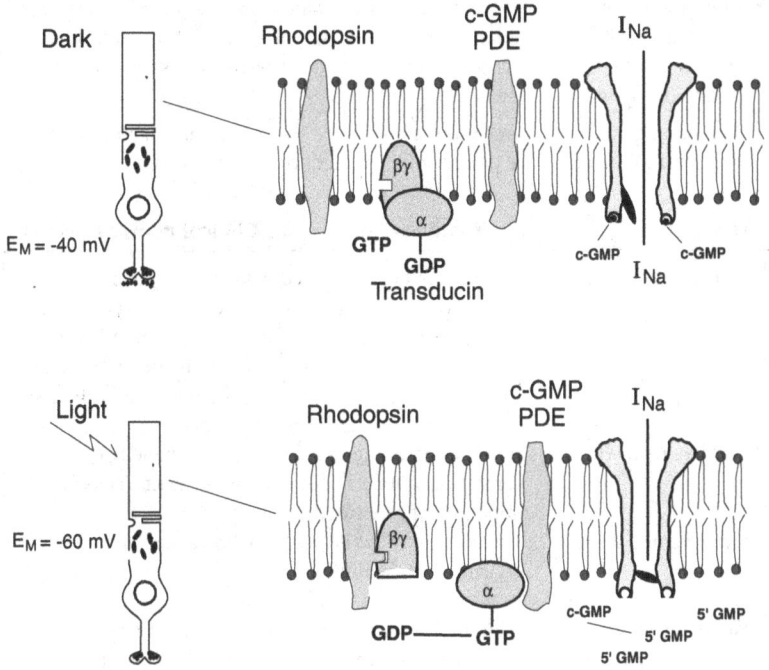

- Spinal reflexes provide rapid control of posture and movement. Ia afferents provide proprioceptive information about muscle length and velocity of movement. The sensitivity of the intrafusal muscle fibers is modulated by γ-motoneurons. Ib afferents provide proprioceptive information about strength of muscle contraction.

Name	Stimulus	Receptors	Afferent Fiber	No. of Synapses
Withdrawal reflex	Skin damage or irritation	Pain receptors (nociceptors)	Aδ and C	Polysynaptic
Stretch reflex	Muscle stretch	Intrafusal muscle fibers within muscle spindle	Ia	Monosynaptic
Lengthening reflex	Muscle contraction	Golgi tendon organ	Ib	Disynaptic

- Movement is initiated by the motor cortex. Motor commands reach the spinal cord through the pyramidal system (corticospinal tract) and the nonpyramidal system (corticoreticular and corticovestibular pathways). Lesions to the nonpyramidal system cause spasticity.
- The basal ganglia and cerebellum assist in the control of movement. Lesions to the basal ganglia produce paucity of movement or uncontrolled movements.

Disease	Possible Cause	Clinical Manifestations
Tardive dyskinesia	Dopamine antagonists used to treat psychotic diseases	Rapid, irregular movements (chorea) or slow, writhing movements (athetosis) of the face, mouth, and limbs
Parkinson's disease	Degeneration of the substantia nigra dopaminergic neurons	Tremor, paucity of movements, cogwheel rigidity
Huntington's disease	Degeneration of GABA-ergic neurons with the striatum	Chorea, ataxia, dementia
Hemiballism	Lesion of the contralateral subthalamic nucleus	Sudden flinging movements of the proximal limbs

- Lesions to the cerebellum produce uncoordinated movements.
- The vestibular system provides information about the position and movement of the head, coordinates head and eye movements, and initiates reflexes that keep the head and body erect. Lesions to the vestibular system result in loss of balance and nystagmus.
- The cortex is responsible for cognition, language, emotions, and motivation.

ENDOCRINE PHYSIOLOGY

- Most of the hormones secreted by the endocrine glands are controlled by the endocrine hypothalamus and pituitary gland. The posterior pituitary gland is divided into a posterior and an anterior lobe. The pituitary gland contains the axons of neurons located within the paraventricular and supraoptic nuclei of the hypothalamus. These hypothalamic neurons synthesize oxytocin (responsible for milk ejection and uterine contraction) and ADH (increases water permeability of renal collecting ducts). The anterior hypothalamus secretes six major hormones.

The synthesis and release of these hormones are controlled by the hypothalamus. The hypothalamic releasing hormones are secreted into the median eminence from which they enter the capillary plexus that coalesces to form the anterior pituitary portal veins. These veins form another capillary plexus within the pituitary from which the releasing factors diffuse to the pituitary endocrine cells.

- Thyroid hormone increases oxygen consumption and, therefore, the basal metabolic rate (BMR) by increasing the synthesis and activity of Na^+-K^+-ATPase. It acts synergistically with growth hormone to promote bone growth. Thyroid hormone is essential for proper development of the nervous system in newborn infants and for normal cognitive function in adults.

Pituitary Hormone	Hypothalamic HormoneAffecting Pituitary Hormone	Major Action of Pituitary Hormone
Thyroid-stimulating hormone (TSH)	Thyroid-releasing hormone (TRH)	Stimulates thyroid hormone synthesis and secretion
Adrenocorticotrophic hormone	Corticotropin-releasing hormone (CRH)	Stimulates adrenocortical hormone (ACTH) secretion
Growth hormone	Growth hormone–releasing hormone	Synthesis of somatomedins by liver, which, in turn, stimulate protein synthesis, organ and bone growth
	Growth hormone inhibitory hormone (somatostatin)	
Follicle-stimulating hormone (FSH)	Gonadotropin-releasing hormone (GnRH)	Spermatogenesis (males) Estradiol synthesis (females)
Luteinizing hormone (LH)	Gonadotropin-releasing hormone (GnRH)	Testosterone synthesis (males) Ovulation (females)
Prolactin	Prolactin-inhibiting factor (dopamine) Thyroid-releasing hormone (TRH)	Breast development and milk production

- Adrenocortical steroid hormones (glucocorticoids and adrenocorticoids) are synthesized from cholesterol. Glucocorticoids promote gluconeogenesis, inhibit the inflammatory immune responses, and are essential for the vasoconstrictive action of catecholamines.
- Insulin and glucagon are secreted by the pancreas. Insulin affects carbohydrate, lipid, and protein metabolism in adipose, liver, and muscle tissues. Insulin is secreted by the β cells of the islets of Langerhans. Insulin increases the entry of glucose, amino acids, and fatty acids into cells. It promotes the storage of these metabolites and inhibits their synthesis and mobilization. Glucagon is secreted by the α cells of the islets of Langerhans. Glucagon's effects oppose those of insulin.

NOTES

NOTES

NOTES

NOTES

NOTES

NOTES

NOTES

NOTES

NOTES

NOTES

NOTES

NOTES